D1093291

STEAM IN AFRICA

A.E.DURRANT A.A.JORGENSEN
C.P. LEWIS

STEAM IN AFRICA

C.STRUIK PUBLISHERS
CAPE TOWN

5918

C. STRUIK (PTY) LTD,
OSWALD PIROW STREET, FORESHORE, CAPE TOWN

FIRST EDITION 1981

DESIGN AND TYPOGRAPHY: W. REINDERS, CAPE TOWN
PHOTOSET AND LITHOGRAPHIC REPRODUCTION:
HIRT & CARTER (PTY) LTD, CAPE TOWN
PRINTED AND BOUND: NATIONAL BOOK PRINTERS, GOODWOOD, CAPE

ISBN 0 86977 139 6

CONTENTS

No
268

PREFACE AND ACKNOWLEDGEMENTS

Steam in Africa is a broad outline of African railways and their steam locomotives, from their various beginnings to the present day or to the end of steam. It would be impossible to give full coverage of the railways of almost 40 countries in one book, and limitations of space have meant that many details have had to be omitted, including locomotive dimensions, numbers, renumberings, etc. Where further information is available, it may be found in the works listed in our bibliography.

On a continent with a history as turbulent as Africa's there have been numerous changes in names, boundaries and control during the present century. Where a country has changed names, we have used the name current at the time, for example we have referred to locomotives which were built for the Gold Coast in 1910 but run in Ghana today. Metric weights and measures have been used throughout, but imperial units have sometimes been included for the sake of clarity.

In gathering material in so vast a field, we have asked many people for help, and the response has been astonishingly generous.

Our thanks go to Peter Bagshawe, Hugh Ballantyne, E. D. Brant, Mr Brennan of the American Consulate in Johannesburg, Collector's Treasury in Johannesburg, Eric Conradie of the SAR library, Marc Dahlström, Nils Huxtable, Muriel Macey of the Kimberley Public Library, David and Keith Patience, J. H. Price, the *Railway Gazette,* Dr P. Ransome-Wallis, C. S. Small, Frank Stenval, Ted Talbot, *La Vie du Rail,* Patrick B. Whitehouse, Alan Wild of the Bournemouth Railway Club, and Jeremy Wiseman for their help with information and in tracing photographs. We are grateful to all those who contributed photographs (see credits, p. 205), which have greatly helped re-create the atmosphere of steam in Africa.

Particular thanks go to officials of the Zambian Railways, to Arthur Kemp, John Suckling, Ben van der Linde, Ron Wiseman and Bill Young of the National Railways of Zimbabwe, and to Ken du Toit and 'Blackie' Swart of the SAR, for making possible the Victoria Falls photographs.

Finally, we thank our wives – Christine, Judy and Melly – for their encouragement and for their practical assistance with typing, map drawing and research.

(Previous pages)

1. **One of Africa's most impressive natural spectacles, the Victoria Falls and gorge, combines with one of man's more stunning engineering achievements to embody the spirit of Africa's railways. A northbound 20th Class Garratt crosses the bridge which spans the Zambezi River.**

2. **Crossing the Little Karoo in South Africa, the Port Elizabeth-Cape Town passenger train startles a family of ostriches.**

3. **A verdant green Kenyan landscape contrasts with the striking red of an East African Railways Class 59 Garratt near Kibwezi.**

4. **A blue-painted Pacific fronts a train of ivory-coloured coaches crossing the endless plains of southern Sudan.**

5 (Overleaf). **As dawn approaches, this Moçambique Railways Atlantic – the last of a noble breed – prepares to whisk passengers out of Lumbo along a westward track.**

INTRODUCTION

In Africa today one is more likely to see a caged than a wild leopard, and a diesel train rather than a steam locomotive, for the latter, even more than the leopard, has joined the list of Africa's endangered species.

Where once throughout the continent widely differing steam engines could be seen, in 1980 only 11 of the 30 railway-operating countries still work steam. Many a rare and exotic type is extinct, such as the express passenger Garratts of Algeria, the main-line Sentinels of Egypt, the 3-cylinder Mountains of Nigeria, the rack tanks of Angola and the modified Fairlies of South Africa.

While many types have been withdrawn through bureaucratic whim, the rule of the survival of the fittest has largely prevailed – a basic natural law which has applied to Africa's locomotives as well as its animals. Where an engine was well adapted to its surroundings it survived, and its success encouraged the building of similar types. That multi-jointed creature, the Garratt, took to the African scene as if it were indigenous. Over the years it worked in no fewer than 20 countries, generally making its home in the more hilly and mountainous areas and holding fast in a world of rapid technological change; in Zimbabwe it has even made a comeback.

Over the past 125 years some 13 000 steam locomotives have operated on the African continent – about 2 000 on metre gauge, 2 300 on 1 435 mm, 6 500 on 1 067 mm and the balance on narrower gauge – a meagre figure to compare with Europe and the USA. Britain alone had more than 20 000 in service at one time. But what Africa lacked in quantity it made up for in variety. As with its locomotives, so Africa's railways are delightfully varied, reflecting the character not only of the lands where they run, but also the half dozen European countries which sought 'a place in the sun' in the second half of the 19th century, and whose railwaymen built and operated systems similar to those they had known at home.

British enterprise gave Africa its first railway – from Cairo to Alexandria in 1854 – as well as those in southern Africa, Kenya, Uganda and the Sudan.

Because of this it is easy to over-estimate the influence of Cecil John Rhodes on the railways of Africa. His dream of a railway from the Cape to Cairo certainly captured current popular imagination, but railways were already established and thriving in Egypt, various French colonies and South Africa before he became an active force on the continent, though it is true that the railways of Zimbabwe and Zambia owe their existence to him.

Spain gained only a toe-hold, with a few kilometres in the northernmost tip of Spanish Morocco and a short rack railway on the island of Fernando Po.

The Italians built extensively in Ethiopia and Eritrea and to a lesser extent in Libya and Somaliland. With characteristic flair they constructed railways over and through mountain ranges rather than around them; their lines distinguishable by high viaducts, long curving tunnels and similar heroic feats of railway engineering.

Germany's role in Africa's railway development ended during the 1914-18 war and was limited to the 600-mm and 1 067-mm lines of German South West Africa (Namibia) and the metre-gauge lines of Tanganyika (Tanzania), Kamerun (Cameroon) and Togoland.

Belgium built important lines in the Congo, and was also the first to apply 25-KV AC electrification to an African main line.

Like those of other colonial powers, Portugal's railways were built mainly to tap the resources of the interior. Hence the isolated character of the lines of Angola and Moçambique, and the fascinating variety of locomotives which ran on them.

France, after Britain, had the greatest colonial mileage in Africa. The French invariably built, and equipped, their colonial lines to those standards they applied at home. By the late 1940s they had virtually eliminated steam on desert lines and, considering steam outmoded, had dieselized all remaining lines by the late 1950s.

Africa is the continent of the narrow-gauge railway – no broad-gauge lines exist, although a 5′ 6″ (1 676 mm) military line in Abyssinia was one of the continent's earliest. Of eight gauges operated, the international standard gauge (1 435 mm) is restricted to 13 500 kilometres in the north and 83% are narrower than 'standard'. Fairly ubiquitous (though absent in the south) is metre gauge, with some 14 600 kilometres in operation. The most common is Cape gauge (1 067 mm) and, as some 60% (47 750 km) of Africa's railways were built to this gauge, it deserves recognition as Africa's 'standard gauge'.

Today more than 80 000 route kilometres, just over half of which have international connections, exist in some 30 countries. A disproportionate share is concentrated in the south, where eight states (comprising only 37% of Africa's land surface) have 60% of the total route kilometrage. Similarly, these eight states operate 73% of the locomotives and move 80% of all the continent's rail traffic. In four of these eight southern states the greatest concentration of steam is still to be found – a large number of endangered species can be seen, and it is possible to imagine yourself for a moment in the Africa of old.

But no stretch of the imagination can carry one back in the way that a photograph can, and that is why we have compiled this book – a record of the diversity of steam that once flourished throughout Africa.

812

NORTH AFRICA

1 STEAM IN THE MAGHREB

Confidently the Algerian driver notched her back from 15% to 10% and, used to this type of engine and its capacity for speed, watched with only mild interest as the needle moved up to 135, 136 and then to 137 km/h, where it stayed for a few kilometres as the unfamiliar countryside between La Falaise and Ailly Sur Noye flashed by. His only worry must have been that he did not know the route and he was probably somewhat reassured when the French pilot proved his vigilance by indicating that they should slow down to 30 km/h for the perway slack at Boves.

The Frenchmen in the cab were also not particularly interested in the speedometer needle, for such speeds were as commonplace as garlic to crews on this main line of the old Nord company – though not, of course, with a Garratt. Yet a little interest, even jubilation, would not have been out of place, for they had just set a world speed record for articulated locomotives, which to this day remains unbroken.

This test run of March 23, 1937, was intended to establish that the new engine, built for Algeria by the Société Franco-Belge in association with Beyer Peacock, could sustain high speed. On leaving Paris, the crew's orders had been simply to bring this, the regular express, into Calais on time. They arrived two minutes early, no doubt pleased that they had fulfilled their brief.

Were they ever told that they were record breakers? Even the Beyer Peacock official history says no more than 'the maximum speed on the run did not exceed 82 mph'.

The thoroughness and attention to detail in the design and testing of the class BT express passenger Garratts of Algeria were typical of the unskimping approach of the French to the railways of their overseas territories. Wherever they built railways in Africa, the French built substantially and they built well.

Long before World War II, the metre-gauge network of Tunisia allowed 110-km/h running by its SACM Pacifics – the highest permissible speed of any narrow-gauge railway anywhere. The biggest concentration of Mallets on the African continent was for the phosphate trains of the Bône-Guelma Company and the intensive commuter services of Tunis. As early as 1927 the first electrification in the Maghreb was opened to carry the phosphate traffic down to Casablanca. The voltage chosen was 3 000 DC, at a time when the railways of mainland France were still restricted to 1 500 DC.

In Arabic 'Maghreb' means 'west', and Morocco, Algeria and Tunisia are, indeed, the far west of the Arab world.

At various times the original Berbers of all three countries were conquered by Phoenician, Roman, Arab and French settlers. Each race has left its mark. The Phoenicians left their alphabet and the tumbled pillars of Carthage; the Romans left ruins of roads, aqueducts and cities; the Arabs left their religion, language and many beautiful mosques and palaces; and the French left a system of government and their railways. Although the railways traverse exotic countryside, they are typically French in construction, rollingstock and signalling – even language, for the official language of the railways remains French – more than 20 years after independence.

6. **A venerable 0-6-0 built by Fives Lille in 1881, No. 113 of the old Algerian Eastern Railway, poses on a viaduct over the Oued Tiater on the Algiers-Constantine main line. More European than African, this engine is typical of latter-day 19th century freight locomotives, known in France as the 'Bourbonnais' type, and in Africa only established on the standard-gauge systems of the north.**

ALGERIA

At the extreme northern and southern tips of Africa, a geological phenomenon exerted considerable influence on the building of railways. Here, and nowhere in between, occur folded mountains for several hundred kilometres inland from the coast, almost as if giant forces had taken the land mass and compressed it in a vice, crumbling the ends. French and British railway engineers – in Algeria and South Africa respectively – found different ways to surmount these rock-girt barriers. In Algeria railways are characterized by sustained gradients of great severity, so that nearly all locomotives were built with substantial adhesion in relation to their overall mass.

The Atlas mountains form a huge wall, shielding less than one-sixth of the surface area of Algeria from the menacing sands of the Sahara. Range upon range of parallel chains of mountains with inter-connecting spurs enclose a series of fertile basins, which increase in altitude and aridity, south from the Mediterranean coast. The first major range is the Maritime – or Tellian – Atlas, north of which lies one of the few regions of arable land, yielding various kinds of grain. Between the Maritime and the higher Saharan Atlas a series of plateaux, rimmed by spurs of the main ranges, are the home of fruit and esparto grass. The whole region is drained by rivers, nearly all short, flowing impatiently towards the Mediterranean through deep valleys and narrow defiles. South of the Saharan Atlas the desert yields important minerals such as oil, iron, manganese, phosphates and coal.

From 1858 onwards four categories of railways–trunk routes, mineral lines, so-called 'penetration railways' and roadside tramways–were knitted into this tortured topography, gradually taking shape as the Société Nationale des Chemins de Fer Algériens (SNCFA), the present-day railway network of Algeria.

Of the trunk routes, the first section, from Algiers to Blida, was begun by the military in 1858 and completed by a private company in 1862. Soon after this, the newly-formed Paris, Lyon and Mediterranean Company (PLM) took over its operation as well as the further construction of the main line to Oran. The French company was generous in its treatment of its colonial offshoot, providing motive power in the shape of its redoubtable 0-6-0 designs of 1856-7, some of which, as SNCFA classes 3B, 3E and 3F, were destined to last almost until the end of steam on the Algerian standard gauge.

Likewise, the PLM civil engineers seem to have looked on this as a colonial railway in location only, and engaged in heavy engineering on a large scale. The worst obstacle was the mountain section for 50 km west of Blida, culminating in the 2,3-km Atlas tunnel at an altitude of 500 m, near Miliana Marguerite, for half a century the longest tunnel in Africa.

When he visited Algeria in 1865 Emperor Napoleon III must have been impressed with this 'Imperial Main Line' for he urged its rapid development.

Algiers to Oran (422 km) was opened to traffic in 1871, while the previous year the 87 km from the port of Philippeville (now Skikda) to Constantine had been completed. More 0-6-0s arrived and the first of the highly successful 0-8-0 designs (SNCFA classes 4A and 4C) were put to work, presumably on the 1-in-80 grades up from Philippeville and on the mountain section of the main line to Oran.

The next stage of the grand trunk route, from Bône (now Annaba) to Guelma, was completed in 1877 by the famous engineering firm Société des Batignolles whose subsidiary, 'Chemins de Fer Bône-Guelma et Prolonguements' (B-G) took over operation of the railway and further construction to Constantine and, via Souk Ahras, to Ghardimaou on the Tunisian frontier. Here, in 1884, the link was made with B-G's line from Tunis, permitting international running between Algeria and Tunisia for the first time.

Operations were begun with 0-6-0s by Batignolles and, by 1883, 39 of these covered the 550 km of the company's main lines. Although they were assisted by 18 Batignolles 0-6-0Ts for shunting and banking, 'train services must have been pretty sparse' according to P.M. Kalla Bishop.

The Algerian Eastern was floated to link Constantine and Algiers, thus connecting the two isolated sections of the PLM. Construction started westward in 1879 and encountered some of the most rugged territory on the Algerian standard gauge, crossing the Tellian Atlas at an altitude of 1 088 m before descending on long, twisting 1-in-50 – and steeper – grades towards Algiers. Completed in 1886, it was an important milestone, providing direct rail communication between the capital cities of Algeria and Tunisia.

Services of the Algerian Eastern commenced with 47 0-6-0s supplied by SACM and Fives Lille. Like the ex-PLM 0-6-0s they survived almost until the end of steam in Algeria, as SNCFA classes 3A and 3C.

The fourth company involved in the building of the main trunk route was the Algerian Western. Construction started westwards in 1877, and by 1890 had reached Tlemcen, the last major town before the Moroccan border. At this stage, with the previously built 'branches' to Ras-el-Ma and Aĩn Témouchent, the Algerian Western could provide services over some 300 km of railway, including the 164 km of main line from Oran to Tlemcen. Here too, the service must have been 'pretty sparse', for the only motive power was provided by 26 0-6-0s. They were built by Fives Lille and SACM and became SNCFA classes 3L and 3M.

By 1910 the Algerian Western had been extended through mountainous terrain over the 835-m summit west of Tlemcen to the Moroccan border. The line was connected with the Moroccan system at Oujda in 1916, when a 15-km narrow-gauge military line was converted to standard gauge.

At the turn of the century the four companies with main line operations provided a barely adequate service. Of the four, the PLM's service was best, and the company was the most financially sound. While all were subsidized by the state, the subsidy of the other three was calculated per km of track in such a way that it did not pay them to attract custom; freight rates were high

7

and passenger schedules slow. According to a report to the Algiers Chamber of Commerce in 1887 – quoted by E.D. Brant in *Railways of North Africa* – it was cheaper to bring grain to Algiers by sea from India than by rail from farms 300 km away.

Eventually, after several years of merely pressing the companies to improve services, the state took them over; the Algerian Eastern and Bône-Guelma by World War I, and the Algerian Western and PLM in 1921. It was agreed that the PLM would operate its old main line from Algiers to Oran and all the lines of the Algerian Western, while the state would operate all lines east of Algiers, including the PLM's old line from Constantine to Phillippeville. This led, in effect, to two separate standard-gauge systems.

In the first decade of this century the motive power was as uninspired as the train service. The Algerian Western operated 20 insipid Moguls and hordes of the same standard French-designed de Glehn compound 4-6-0s as were to be found in France, Spain and Portugal at the time. The Algerian Eastern also acquired 4-6-0s for their passenger services – three from Five Lilles in 1904, followed by 47 from the same firm and SACM in 1911. These became SNCFA classes 230F and D. As usual, the PLM tried harder.

They used their four-cylinder de Glehn compound 230As with 'streamlining' behind the chimney and dome, windcutter cab fronts and cone-shaped smoke-box doors. As the best overall speed between Algiers and Oran was 39 km/h, the worth of these embellishments was probably slight.

The most interesting locomotives arrived in Algeria between the two World Wars. Soon after the reshuffle of 1921, the Schneider 2-10-0s of classes 150A and 150B were introduced. These were based on the B-G's Société Alsacienne design of 1910 for its Tunisian lines and were intended to work the strenuous ore conveyor between Oued Keberit and Bône. Schneider also delivered 15 fine 4-8-2s (SNCFA class 241A) for passenger work on the former Bône-Guelma lines. These were equipped with counterpressure braking – steam's equivalent of the dynamic brake – which was ideal for the 1-in-38,5 drop off the Medjerda mountains down to Bône, Souk Ahras and Ghardimaou on the Tunisian frontier. The engines were unlike anything previously seen in Algeria. American in concept, with two simple cylinders and bar frames, though chunky in appearance, they were similar in size and power to South Africa's class 19.

Meanwhile, the PLM was stocking lines west with yet more 4-6-0s class 230A and an enlarged version of it with heavier axleload – class 230C – equivalent to the same classes on the mainland PLM. They also acquired several 2-8-0s for freight service, including ex-World War I US Army Pershing 2-8-0s (SNCFA class 140G) and Batignolles 2-8-0s of class 140C.

Over the eastern main lines 30 0-8-0s of the Prussian Railways G8 type were handy for heavy freight work. Another Prussian design, the legendary three-cylinder conjugate-geared G12, arrived from Franco-Belge in 1928 and, as SNCFA class 150C, became the staple heavy freight engine on the main lines until the end of steam.

By the 1930s the PLM faced increasing difficulty in handling the Algiers-Oran passenger services with its aging 4-6-0s. They were used in relays and any load of more than 256 tonnes had to be doubleheaded, while on the slow night trains of 500 tonnes and more, a third engine helped on the mountain section up to the Atlas tunnel. The PLM wanted a machine which could work through to Oran without change and handle up to 500 tonnes on its own.

This set the stage for the début of one of the most striking locomotive classes of all time – the express passenger Garratts of Algeria.

The prototype, no. 231-132AT1, was built by Franco-Belge in 1932 and, after extensive trials on the PLM in France, was tested between Algiers and Oran. While these tests were being carried out, another milestone in Algerian railway history was reached when the State took over PLM operations and placed the two standard-gauge systems under joint management of the SNCFA.

The new management tried their newly-acquired passenger engine on the much more difficult Algiers-Constantine main line with such impressive results that another 12 Garratts of improved design were ordered. These were delivered by Franco-Belge during 1936. One of them, no. 231-132BT11, made the record-breaking run described at the beginning of this chapter.

Where the prototype had Beyer Peacock's self-trimming rotating bunker for hand firing, single Kylchap exhaust (later changed to double Kylchap) and Walschaert valve gear (later changed to Cossart), the production batch had double Kylchap exhaust, electrically-reversed Cossart valve gear and mechanical lubrication, and they were strikingly streamlined. Classified 231-132 BT, they had considerably more power than No. 231-132AT1, by virtue of increased boiler pressure of 20 kg/cm² (284 lb/sq in), the highest ever applied to a Garratt. With their starting tractive effort of 30 000 kg (66 000 lbs), they were the most powerful express passenger locomotives ever operated outside the United States.

Subsequently, Franco-Belge supplied 17 more. As the last entered service in 1941, at least some of these must have been shipped from France only a jump ahead of the German invasion. Before war reached Algeria several modifications were made: all were equipped with mechanical stokers as it had become obvious that, even with two firemen, optimum performance could not be sustained on the long grades; initial troubles with the Cossart gear were ironed out when special arrangements were made for its maintenance; and some were converted to oil fuel.

The Garratts were an undoubted success in peacetime operation, spectacularly improving passenger services. In 1937 the Algiers-Oran service was accelerated from nine hours to just under seven hours for the 422 km, including 19 intermediate stops with 450-tonne maximum load for a single Garratt. Today's diesels take five-and-a-quarter hours, with only nine stops and a mere 200-tonne load. To keep time the Garratts had to run at 120 km/h over the easier stretches – the fastest scheduled trains ever to run in Africa.

According to Baron Vuillet, this service was maintained with ease. In 1937 he recorded no. 231-132BT1 holding 65 km/h minimum on the 1-in-45 equivalent grades up to Miliana, with boiler pressure and water level constant and a 346-tonne train. On the Algiers-Constantine line the improvement was even more spectacular, the running time being cut by a remarkable four hours to eight-and-a-half hours for the 464 km, including all stops, on a line which rises to 1 088 m on 1-in-38 gradients.

The earlier intensive testing in France produced some impressive statistics. They were free-running and fast; they could produce 3 000 drawbar horsepower and could sustain 3 600 IHP; and they were reasonably economical, consuming 1,8 kg (3,7 lb) of coal per DBHP-hour when working flat out – slightly above 25% more than a Chapelon 4-8-0 doing the same work – but the latter was, of course, exceptionally light on coal.

The Garratts came to Algeria with impeccable credentials and, after teething problems, gave good service. Yet, within ten years a design which should have given two or even three decades of service was retired in disgrace. Why were machines, which seemingly had everything going for them, cut down in their prime?

The answer might lie in their many points of deviation from current locomotive practice. It could not have been the Garratt principle itself, for by the 1930s Beyer Peacock's patent had been proved thoroughly; mechanical lubrication had already established itself in America and would become standard practice for modern steam; the application of American mechanical stokers to Garratts was not unique – South Africa had nearly 200 large Garratts so equipped, which were trouble-free. The electrically-reversed Cossart gear was, perhaps, a little adventurous for African conditions, but maintenance standards in Algeria, at least until World War II, were certainly on a par with those in mainland France, and, in its only other major application – to the 70 Nord 141TC suburban engines of Paris – the Cossart gear proved able to withstand more than 40 years of hammering in one of the toughest steam commuter services in Europe.

8

7. An impressive line-up of 0-6-0s, staple main-line power of the Algerian Eastern Railway, at Sidi Mabrouck about the turn of the century.
8. A Class 230A de Glehn compound locomotive 'A bec', of the PLM, about to leave Oran with a rapide for Algiers before World War I. The seemingly incompatible French rollingstock and Arabian architecture blend delightfully into this north African landscape, dominated by a hill-top Moorish castle.
9. The station could have been transported brick for brick from mainland France. The wagons and coaches are quite continental, while the locomotive, No. 84 of the Bône-Guelma company, is a Baldwin graduate of 1899, originally a 2-6-0 but rebuilt early in the 20th century as a 2-6-2T. But the Medjerda mountains in the background are stubbornly African and form the major obstacle between Souk Ahras and the rest of Algeria.

If the design was not a lemon, what caused these engines to fail? No steam engine likes water carry-over caused by priming. The Cossart gear, with its two pairs of vertical piston valves at each end of the cylinders, was especially sensitive in this respect, so it became necessary to empty and refill the boiler after each run and to wash out after each round trip between Algiers and Oran or Constantine. In spite of this, Baron Vuillet records that the engines required careful handling to avoid priming at the end of a long run. Correct water treatment would have minimized priming and carry-over of water and the resulting damage to the 'delicate' valve gear; and on better water the engines could have run for four weeks or longer without being washed out, which would undoubtedly have improved their unimpressive monthly average of 5 000 km recorded in 1938.

The bad Algerian water seems the major cause of the Garratt's problems.

9

10. **No. 230C12 suspended from the overhead crane in the workshops at Algiers. These four-cylinder compounds were the last heavy passenger power with which the PLM stocked its Algiers-Oran main line in bulk, and were well-suited to the steeply-graded route when handling trains of up to 256 tonnes. On a continent known for narrow-gauge railways, Algeria operated more than 530 standard-gauge steam locomotives.**

In peacetime, with careful handling, they had been able to cope, but under wartime conditions such handling was not always possible. When in 1942, the Allies tackled Rommel's army from the rear, British and American troops landed in Morocco and Algeria *en masse* bringing with them railwaymen and railroaders – the railwaymen working all lines east of Algiers and the railroaders all lines west. And, when they were needed most, the Garratts began to fall down on the job.

P.M. Kalla Bishop graphically describes the problems British Army fitters experienced with these locomotives: 'Complicated failures were always occurring; the Cossart valve gear was reversed by electric motors and the batteries were always mysteriously going flat. The inability to reverse the locomotive after the batteries had gone flat, or water got into the circuitry, was tiresome and the rotary valve gear did not take kindly to small-size piston valves full of water.'

Strangely, the normally-reliable mechanical stokers also gave trouble, perhaps because of a wartime scarcity of spare parts, and most of the Garratts reverted to hand firing – with three Algerian firemen. Gradually, as major failures occurred, the Garratts were dumped out of service to be replaced by numerous new US Army 2-8-0s which were no match for a Garratt in power but were much more reliable.

The main artery to the Tunisian front was the line to Constantine on which the three-cylindered 150Cs proved their worth. British fitters of the LNER would have been familiar with their conjugate gear, and these well-maintained machines handled a daily service of 11 trains each way supported by the two-cylindered 150Bs and the older 2-8-0s, 4-6-0s and 0-8-0s, the latter presumably relegated to banking duties.

The militarily important line from Philippeville to Constantine was worked by ex-US Army World War I Pershing 2-8-0s, followed later by the new World War II 2-8-0s, and 151ATs from the mines at Ouenza were used for banking. On the more easily-graded (1-in-100) route eastward from Constantine towards Bône, Pershing 2-8-0s and ex-PLM four-cylinder compound 230Cs handled the traffic.

From Bône, the route eastward used the electrified Ouenza mineral line as far as Souk Ahras from where 150As and 150Bs banked the electric-hauled ore trains, as they also did up the strenuous section from the Tunisian border at Ghardimaou.

Shunting was mainly handled by the 80-year-old PLM 0-6-0s, though the British War Department sent six 'Dean Goods' early in 1943 to lend a hand for a few months.

Although the PLM had experimented with main-line diesels simultaneously with their first Garratt, serious application of diesel-electric power occurred only after the war with the delivery of 60 Baldwin units between 1946 and 1948. The 25 for express passenger service, geared for 130 km/h, put paid to the Garratts. As early as 1948 steam was handling only 20% of the reduced post-war traffic, and by 1958 all standard-gauge steam working had ceased. The early demise of steam is thus largely attributable to the bad Algerian water and indifferent coal, mined more than 1 000 km from the main cities, coupled with continuing discoveries of high-grade oil in the Sahara.

When, soon before the war, Algerian lines of all gauges were merged, SNCFA embraced the three types of railway other than trunk routes – the mineral lines, 'penetration' railways and roadside tramways.

The first mineral railway in Algeria was the 38 km metre-gauge line from Bône to the iron ore deposits at Ain Mokra, opened in 1862. By 1868 small 0-6-0Ts moved a yearly 200 000 tonnes of ore down to the harbour. In 1904 the line was extended to St. Charles on the standard gauge to Phillippeville and 12 0-4-4-0Ts were supplied by SACM. These presumably worked until 1954 when the line was widened for the new standard-gauge link between Constantine and Bône, replacing the original route to the east via El Kroub and Guelma.

Today's most heavily used line in Algeria, from Annaba to Tebessa, is equipped with 3KV electrification and CTC. But its southern half, south of Souk Ahras, began life in 1888 as a humble 128 km metre-gauge penetration line belonging to the Bône-Guelma Company. The discovery of phosphate near Tebessa and, later, of iron ore transformed its significance in Algeria's rail network.

Originally, 12 Batignolles 0-6-0Ts of 1886 provided motive power, but these would have lacked the power to handle the phosphate traffic and in 1895 the company provided eight Batignolles 0-4-4-0Ts – a big improvement in mountainous terrain.

Immediately before World War I, iron ore deposits were discovered at Ouenza, about half way down the Tebessa line near Oued Keberit. Difficulties of ore trans-shipment led to conversion to standard gauge all the way from Souk Ahras to Ouenza in 1922, the original line to Tebessa remaining metre gauge south of Oued Keberit.

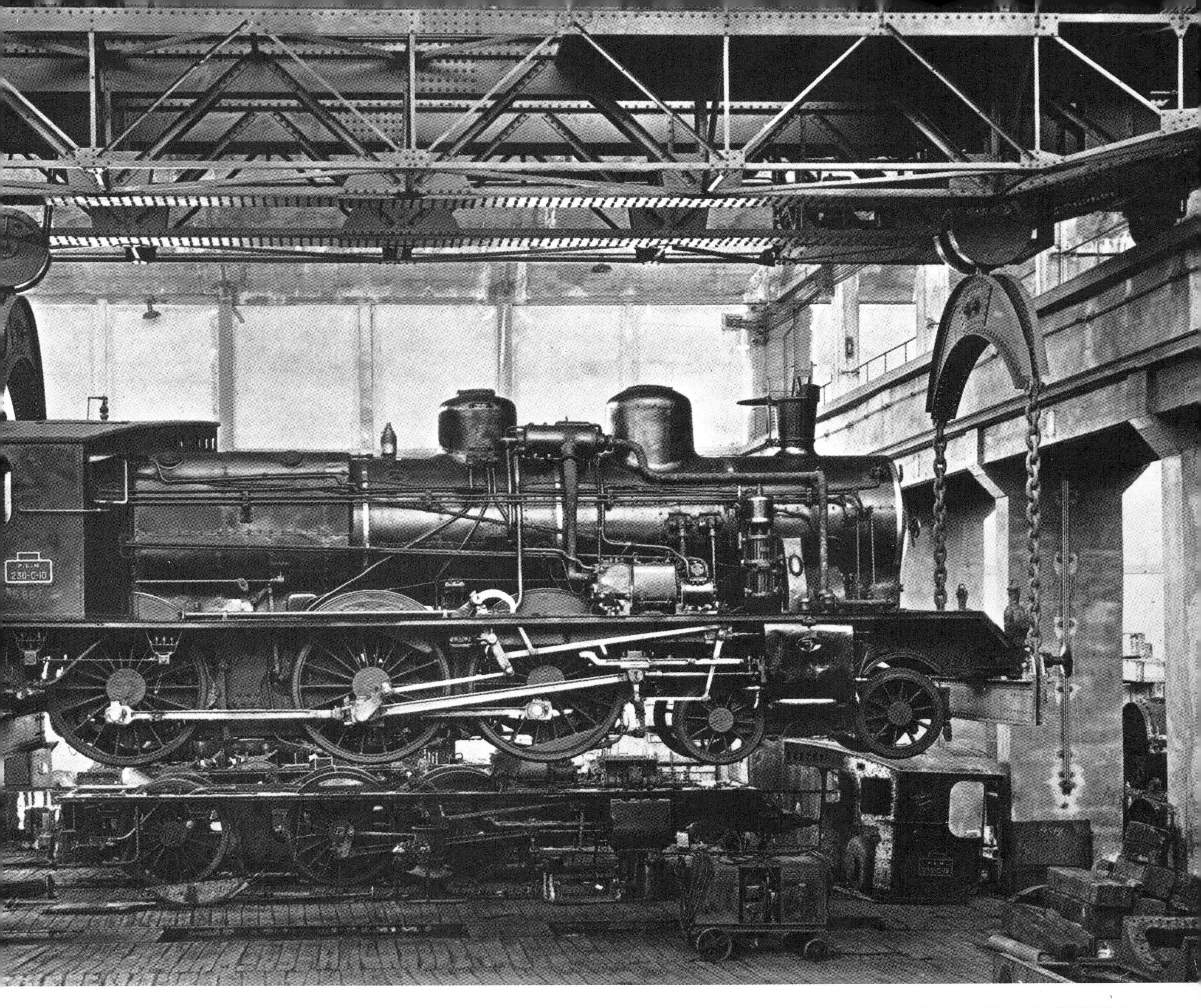

11. **Class 241As were ordered for the Algerian State Railways and were equipped with counter-pressure braking – steam's equivalent of the dynamic brake. They were very successful and ran for more than 30 years after their introduction in 1921.**

12. **The narrow-firebox version of the legendary Prussian G12 became SNCFA Class 150C and was the standard main-line freight engine from 1928 until 1948.**

BEDEL C

13
14 15

13. An extremely rare view of No. 231-132BT11 setting off on its record-breaking run from Paris Nord to Calais in March 1937. The most striking locomotives in Algeria, indeed in all Africa, these express-passenger Garratts were built by Franco-Belge and thoroughly tested in France before being shipped to Africa.
14. The first express-passenger Garratt, No. 231-132AT1 on the turntable at Algiers during the mid-1930s. This view shows her as she was rebuilt with Cossart valve gear and double Kylchap exhaust with the chimneys arranged transversely. In this form she became the prototype for the production batches of Class 231-132BT 1-29. This engine was later modified once more with triple chimneys arranged in a clover leaf pattern.
15. Achilles heel of the express Garratts was the electrically-operated Cossart valve gear. The electric contact box for altering the cut-off of valves is clearly visible.

No less than 60 two-cylinder 2-10-0s (SNCFA classes 150A and 150B) were built by Schneider for working the through ore-trains from Oued Keberit to Bône, as well as the steeply-graded Algerian section of the original B-G main line between Souk Ahras and Ghardimaou. The mines at Ouenza had some 2-10-2Ts which eventually came to the SNCFA as class 151AT and banked trains up from Phillippeville in World War II.

The standard double-headed ore drags of 1 120 gross tonnes faced heavy grades northwards from Oued Keberit: first climbing at 1-in-62 for 24 km to 889 m at M'Daourouch, followed by a long descent to Les Tuileries where a third engine would be attached for the 11-km stint – at 1-in-40 and including 13 tunnels – to the summit, nearly 1 000 m above sea level and 8 km north of Souk Ahras. Gasmasks were used by the crews when working through the tunnels, and it must have been particularly uncomfortable on the third engine. Inevitably, electric traction was introduced from Bône to Ouenza in 1933, although the 150As and Bs continued to bank electric ore-trains up from Les Tuileries until after World War II.

South of Oued Keberit, four of the old B-G Mallet tanks continued to ply their loads of phosphates northwards, their work considerably eased in 1931 by the extension across the border to Tebessa of the B-G (the company still operated its Tunisian lines) metre-gauge phosphate line from the port of Goulette to Kalaa-Djerda.

In 1946, to eliminate expensive trans-shipping at the break of gauge, the metre gauge south from Oued Keberit was widened. The line through Tebessa to the phosphate deposits at Le Kouif was electrified in 1952 at about the same time as the international connection from Tebessa via Haïdra for export through Goulette ceased to be used.

From Miliana Margueritta, at the western portal of the Atlas tunnel, a short, steep 600 mm-gauge feeder served the Zaccar iron ore mines about 230 m higher, near the ancient town of Miliana. The first Garratt in Algeria was built by St. Léonard to work this line as early as 1912 and a further two were supplied by Haine St. Pierre during 1936/37.

The last mineral line to be built began life in 1939 as part of an ambitious project to build a railway across the Sahara to Niger. Starting at the port of Nemours (now Ghazaouet) near the Moroccan border, it joined the main international trunk route about 40 km inland and followed it to just west of Oujda, in Morocco, where it branched south on an existing Moroccan route serving the manganese mines at Bou Arfa. From 1942 an extension from Bou Arfa served the coal mines at Bechar and Kenadsa, but when these mines closed, in about 1960, so did the extension. Today only Bou Arfa's manganese finds its outlet to the sea at Ghazaouet.

SNCFA class 4E 0-8-0s (Prussian G8 type) were used during construction, but because of water problems this route was early to dieselize.

Penetration railways

Today Africa's railway map still features numerous isolated lines built by the colonial powers to tap the raw materials of their possessions. The so-called 'penetration railways' of Algeria were no exception, though some were motivated by military strategy too. The lines were built to three gauges – standard, metre and, by far the most important, 1 055 mm.

16

17

18

On the high plateau of Oran an interesting curiosity preceded railway construction. This was a Lartigue monorail, cheaply constructed over difficult terrain and built to transport esparto grass. But, added to the line's inherent disadvantage – poor junction facilities, imbalance of rolling stock and low speeds – was one unique to Algeria: shepherds insisted on removing sections of the raised track so that their flocks could move from one side to the other.

Grazing having put paid to this experiment, the first of the 1 055 mm-gauge lines was laid down in 1874, the same year that construction began on the practically similar gauge at the southern tip of the continent. However, in Algeria the narrow gauge ultimately merely supplemented the standard gauge, feeding the main lines. The Franco-Algerian Company built this first line from the port of Arzew, 40 km east of Oran. Ostensibly the bait was the vast area on the Oran plateau suitable to grow esparto grass, but military considerations were behind the extension southwards, off the plateau towards the Moroccan frontier. In 1906 the line eventually reached Bechar, nearly 800 km from Oran, then the heart of the only known Algerian coalfields.

Franco-Algerian built another isolated penetration line southwards on the heavily-graded track from the port of Mostaganem to Tiaret, completing the 200 km by 1889. Financially exhausted by its over-optimistic projection of receipts from conveying esparto grass and military personnel, the company was forced to hand its lines to the Algerian Western, which sold the embryo narrow-gauge network to the state in 1904. Under state control, an ambitious programme of extension and integration was begun, the first step being to connect the Mostaganem-Tiaret and Arzew-Bechar lines with Oran. The kilometrage of the 1 055-mm lines expanded until 2 200 km were in operation in 1938, when politicians again changed railway history.

The French government decided to co-ordinate the transport system of France, including Algeria, with a view to eliminating 'wasteful' competition.

19

16. A local passenger train of the Algerian Western Railway crosses the Pont des Cascades in the mountainous terrain west of Tlemcen.

17. A typically French narrow-gauge 4-6-0 basks in the sun outside the 1 055 mm-gauge shed at Blida in August 1951. Note the double transverse chimneys of the Garratt in the background.

18. The design for Class 241-142 YAT was based on the Kenya-Ugandan EC1, prepared orginally by Beyer Peacock and updated by the French. It has round-edged tanks, triple headlights, feedwater heaters and airbrake cylinders. Hidden from view is the single Kylchap exhaust (later changed to double Kylchap with transverse chimneys – see picture 17) which made these formidably efficient engines the most successful ever to run on the Algerian narrow gauge. Over the years Algeria had some 280 engines of 1 055 mm-gauge and 74 on metre gauge.

19. These delightful little 600 mm-gauge machines could have been designed by Rowland Emmett. Among the earliest of African Garratts, they were supplied to the Zaccar iron mines in 1912 and 1936.

In Algeria construction of all new narrow-gauge lines ceased, and several existing lines were closed. The 1 055-mm lines were reduced to feeders for the main trunk routes.

The first line-service locomotives bought by Franco-Algerian were 27 0-8-0s, supplied by SACM and Esslingen from 1876 onwards (SNCFA class 4YA). With maximum adhesion, they suited the long pulls – at 1-in-37 to 1-in-50 – to the 1 151-m summit at Tafaroua on the Bechar line some 200 km from the sea. The Mostaganem-Tiaret line apparently started operations with 10 4-6-0s by Fives Lille and SACM for passenger work (SNCFA classes 230YC, 230YL and 230YM) and 45 2-8-0s for freight service by Fives Lille (class 140YA) and Schneider and SACM (class 140YB). The biggest non-articulated engines on the 1 055 mm lines were five 2-10-0s by Franco-Belge, supplied in 1927 (class 150YA). This road power was supported by Franco Algerian's large cast of shunting and banking engines, ranging from 0-6-0Ts to 2-8-0Ts.

The most interesting engines to work the narrow gauge were four 4-8-2+2-8-4 Garratts which the PLM ordered in 1931 for its Blida-Djelfa branch. This branch had been started by the Algerian Western in the 1890s and, after the big re-organization of 1921, became the only narrow-gauge operation of the PLM. Financed by the state, but under PLM supervision, the line was extended during the 1920s to Djelfa, at the foot of the Saharan Atlas. However, a plan to penetrate the range through the Caravan Pass and descend to the desert was not carried out, probably because of the rationalization programme of 1938.

The line as built was, nevertheless, most spectacular. Five kilometres from Blida the ascent of the northern chain of the Atlas involved a climb of over 1 000 m in a distance of 65 km on a ruling gradient of 1-in-40, uncompensated for curves as tight as 120-m radius. Up this climb the Garratts were allowed 360 tonnes, about three times as much as the conventional 2-8-0s. Classified 241-142YAT, the Garratts were an unqualified success. Though based on the proven Beyer Peacock EC1 design for the Kenya-Uganda Railway, they were thoroughly French in detail, with tanks neatly rounded along all edges, stove-pipe chimney, ACFI feed-water heater and three headlights. They were later equipped with transverse double Kylchap exhausts, in line with the prototype PLM main-line Garratt which left the Franco-Belge works almost simultaneously. They showed a fuel saving of 17% over non-articulated types, and gave nearly 30 years' service before diesels replaced them. Their introduction made Algeria one of the only two countries in Africa to run Garratts on three gauges, the other being Belgian Congo.

The other penetration lines were started by the Algerian Eastern. The first, on standard gauge, was built in the 1880s stretching south from the AE main line at El Guerrah near Constantine. It terminated at Biskra, a large oasis at the northern rim of the Sahara, until a 220-km southward extension on metre gauge was built by the state to Touggourt where the bulk of the Algerian date crop is produced. This extension passes through a depression 25 m below sea level.

With the discovery of important oil fields south of Touggourt it was decided to widen the metre gauge from there to Biskra, but by the time this was completed in 1957, all operations had been long-since dieselized.

The old Algerian Eastern 0-6-0s must have worked their hearts out on the 110-km 1-in-62,5 climb up the Saharan Atlas northbound from Biskra – a route which includes six tunnels and the gorge of El Kantara.

In the final decade of the 19th century the Algerian Eastern built a metre-gauge branch reaching south-east from Oued Rhamoun, near Constantine towards Tebessa, reached in 1927.

Initially the line was worked by small 0-6-2Ts, built by SACM (class 31XA), but in 1908 SACM provided eight 0-6-6-0T Mallets. Traffic must have declined in later years, as Brant records that they were converted to 1 055 mm gauge, presumably to operate on the Blida-Djelfa line.

Before the 1938 transport co-ordination decree, extensive systems of roadside tramways of both 1 055 mm and metre gauge operated in Algeria. They resembled the departmental railways of mainland France and had similar motive power. The most important were centred on Algiers, Oran and Bône. The 1 055 mm-gauge system centred on Algiers was operated by Algerian Road Railways (CFRA). The suburban lines of the CFRA were gradually electrified, while country lines were extended to 218 km before World War I. Soon after World War II the last of the non-electrified roadside tramways, those from Oran and Bône were closed.

TUNISIA

As in many other African countries, the first railways in Tunisia were inextricably linked with international politics.

Ahmed Bey of Tunis had borrowed heavily to build himself a sumptuous palace 4 km outside Tunis. The British consul, Richard Wood, seeking to ex-

20

21

22

20. **The last of some 130 standard-gauge steam locomotives to run in Tunisia was this Austerity 0-6-0ST, No. 3-56, a Vulcan product of 1945, shown simmering at Tunis in 1966. In the background is Batignolles metre-gauge 0-4-4-0T No. 2-2-460 built in 1903.**

21. **The last SACM 2-10-0, No. 150-234, in steam at Tunis in 1966.**

22. **Among the last steam locomotives to run in Tunisia were some ancient 2-6-0Ts, rebuilt from 1879 0-6-0Ts in the early 1900s. The engine takes on water, while the sack hanging from the cab-side probably contains liquid refreshment for the crew.**

tend his country's influence in Tunisia, persuaded Ahmed Bey's successor to grant a concession to a British company to build a 20-km railway from the palace into Tunis and on, past the site of the ancient city of Carthage, to La Marsa. E. Pickering built the standard-gauge line which went into service in 1874 with four Sharp Stewart inside cylindered 2-4-0Ts.

Pickering had been responsible, 15 years earlier, for initial construction on the Cape Town-Wellington railway (in what is now South Africa), but he ran into financial difficulties and was displaced from the construction work. His Tunisian venture was also a financial failure and two firms – one Italian, the other French – competed for control of the line, with the Italians finally paying a highly-inflated price for this foothold in Tunisia. They operated the line from 1876 until its take-over by the Bône-Guelma Company in 1898. After it became part of Tunis Tramways in 1905, it was electrified.

Italian ownership of the railway had political repercussions, for French fears that Italy would extend her influence in the region was one of the factors leading to the French occupation of Tunisia in 1881. Wood's scheming was in vain, for in 1875 the British government came to an agreement with France, Britain being allowed to keep newly-won Cyprus in return for a free hand for France in Tunisia.

Just as the enterprising Société de Construction des Batignolles and its able subsidiary, the Bône-Guelma Company, were responsible for the railways of eastern Algeria until well into the 20th century, so they built and operated almost the entire Tunisian network, with the major exception of the metre-gauge phosphate carrier of the Sfax-Gafsa Company.

Batignolles started building westward on standard gauge from its own terminus in Tunis in 1876, intending to link with the B-G's lines in eastern Algeria. Following the fertile valley of the Medjerda, once a granary of the Roman Empire, the construction crews reached Souk-el-Arba in 1878 and Ghardimaou, near the frontier, in 1880. The international connection to Souk Ahras, with 1-in-40-plus gradients and several tunnels along the upper reaches of the Medjerda, took another four years to complete. This was the first link on the 'Imperial Main Line' eventually connecting the capital cities of Algeria, Tunisia and French Morocco. In recent times it lost its glamour when the through-passenger services, with Wagons-Lits dining cars, were discontinued. Through-passenger service between Algeria and Tunisia has since been resumed.

Apart from the 75-km branch to the harbour of Bizerta, no further standard-gauge construction took place in the 19th century, and the standard-gauge network reached its maximum extent in 1922, when the branch to the iron ore mines of Tabarka was completed.

In the same year the State took over ownership of all the Bône-Guelma Company's lines in Tunisia, though operation remained in the hands of the company, which changed its name to 'Compagnie Fermiére des Chemins de Fer Tunisiens' (CFT). Just before independence from France in 1956, the State took over the operation of the lines as well, so ending 80 years of participation by the Bône-Guelma company in Tunisia's economic development.

As Batignolles had provided all the rails, bridging and other material, it is no surprise that they also supplied all the Bône-Guelma's early locomotives. These seem to have been distributed at random between Tunisia and Algeria, but afterwards Tunisia had impressive locomotives long before Algeria did. In 1910, the B-G acquired from SACM 15 two-cylinder 2-10-0s of an advanced design, which became the basis of the Algerian 150A and 150B class of 1920-21. The last survivor was seen in steam at Tunis MPD as late as 1967.

The Bône-Guelma apparently took both its main-line and its surburban passenger services seriously, a tradition subsequently maintained by the CFT and SNCFT. Between 1914 and 1928 the company bought 12 four-cylinder simple Pacifics from SACM for passenger work. These were the last standard-gauge steam locomotives ordered by the Tunisian Railways.

In World War II, six each of the American WD 2-8-0s and the unattractive British WD 0-6-0STs arrived to assist with traffic to the Tunisian front. Two of the 0-6-0STs managed to survive into the 1970s as the last serviceable steam locomotives in Tunisia. Further wartime help came from six Dean Goods 0-6-0s during their temporary stay at Mastouta. These venerable machines had come forward with the British army from Algeria.

23. **The B-G's metre-gauge SACM Pacifics were speedy and efficient. Five were bought by Spain's La Robla railway in 1958 and one of these is shown at Cistierna on the Correo from Bilbao in 1962.**

Tunisian Metre Gauge

In 1895, under the auspices of the B-G, the first rails of what was to become an extensive metre-gauge network were laid from Tunis south towards Nabeul and Sousse on the coast. The following year a start was made on the important mineral line to Kalaa-Djerda near the Algerian frontier.

In 1908 the Sousse-Kairouan line was used as a springboard for a new railway to the phosphate mines of Metlaoui, forming an end-on junction with the Sfax-Gafsa Company's terminus at Henchir Souatir. This railway would have competed directly with the Sfax-Gafsa, so there must have been some agreement between the two companies as to division of traffic.

When it came to motive power for the metre gauge, the B-G again excelled itself, with a fleet of 135 Mallet tanks in four different classes – the largest concentration of Anatole Mallet's patent in Africa. Early metre-gauge power had been a nondescript collection of 0-6-0Ts, 2-6-0Ts and 2-6-0s, which would have been too weak for the type of service envisaged by the B-G for its mineral lines.

The first batch of Mallets comprised eight 0-4-4-0Ts, which Batignolles delivered in 1895, a decade before the railway to Kalaa Djerda was completed, and it is assumed that they spent their early life on the Souk Ahras-Tebessa mineral operation in Algeria. Another 32 0-4-4-0Ts arrived from Batignolles between 1903-06, followed by 65 0-6-6-0Ts by Batignolles, Franco-Belge, Henschel and Schwartzkopff in 1907. These bore the brunt of the mineral traffic until 1920, when Baldwin supplied another 30 0-6-6-0Ts similar in size, but more American in design, with bar frames, and piston valves on the H.P. cylinders. They had split side-tanks, the forward section being mounted on the front engine-unit, thus articulating with it on curves.

From 1926 onwards 34 of the European 0-6-6-0Ts were rebuilt in the works at Sidi Fathallah, emerging as 2-6-6-0Ts with larger boilers and tanks. Soon after the start of dieselization in 1952, 11 of these fine-looking engines were sold to the Utrillas Railway – the first of several B-G metre-gauge types sold to private railways in Spain.

To help with the export of phosphates from the Tebessa region of Algeria, in 1931 the B-G extended the 'Mining Grand Central' from Kalaa Djerda across the border. Though longer than that from Tebessa to Bône, the Tunisian route was much more favourably graded for loaded trains, and involved no break of gauge. It seems that this extension was busy until after the arrival of Rommel's forces, when it faded into disuse. It was probably abandoned soon after the standard gauge to Tebessa had been completed in 1946, but at one time it must have been considered an important link with Algeria because Wagons-Lits were provided on through international trains between Tunis, Kalaa Djerda and Tebessa, and adventurous souls could have travelled behind metre-gauge steam from Tunis to within 30 km of Constantine.

This connection generated so much extra traffic that the B-G deviated from its traditional Mallets and ordered 15 2-10-0s from Franco-Belge. These were delivered in 1930. In 1953 three were sold to the Penarroya-Puertollano railway in Spain, the rest being stored until they were scrapped some time before 1959.

In 1912 the B-G linked up the last two major towns in the south, Sousse and Sfax, creating a through line from Tunis which has always catered predominantly for passengers. For this traffic the B-G relied on 4-6-0s of various classes, having 21 in service by 1908, but suburban trains between Tunis and Hamman-Lif were handled by the 0-4-4-0Ts, similar to the renowned suburban service of Oporto.

24

25

26

27

In 1912 the first batch of five engines of a classic Tunisian type, the metre-gauge Pacifics, arrived from SACM, a further three being delivered in 1928. The entire class is reputed to have pulled Tunis-Sfax trains at up to 110 km/h – the record for speeds regularly attained on narrow gauge, until officially equalled on the South African Railways in 1973. In 1958, after dieselization, five Pacifics were sold to the La Robla railway in Spain, who painted them a wonderful shade of green and ran them until 1965.

As early as 1896 the 'Compagnie des Phosphates de Gafsa' obtained a concession to mine and export the extensive phosphate deposits around Métlaoui near the oasis of Gafsa. Within three years they had built the metre-gauge Sfax-Gafsa railway and later extended it to newer phosphate mines at Redeyef and Henchir Souatir – which, as mentioned earlier in this chapter, was also reached by the B-G's line from Sousse in 1909. Branches were opened to the oasis at Tozeur in 1912 and to the port of Gabes in 1916.

Early motive power on the Sfax-Gafsa line was provided by 17 Corpet Louvet 2-6-0Ts, six Baldwin Moguls and six Corpet Louvet 0-6-Ts. A batch of eight SACM 2-8-0s delivered in 1904-05, improved the position, but the first really effective engines arrived in 1908 at the same time as the first standard 6-wheeled ore wagons. These wagons had a high load/tare ratio and, to make full use of them, SLM supplied, in the same year, the first of a succession of fine 2-10-0s, 36 being delivered by 1925. The final new Sfax-Gafsa steam class was their SLM-built 0-8-0T, four of which arrived in 1913 and four in 1925. Presumably these were used for banking up the 1-in-100 facing loaded eastbound trains for 60 km soon after leaving Métlaoui.

As if anticipating all the fighting that was to take place in Tunisia during

24. **From 1926 the Sidi Fathallah works of the B-G began converting the 0-6-6-0Ts to 2-6-6-0Ts. These were a great success and after the B-G dieselized, many were sold to the Utrillas railway in Spain. One of this batch, No. 204, was seen near Zaragossa in 1965, nearly 40 years after conversion.**

25. **An 0-6-6-0T of the Bône-Guelma Company ex-works in photographic grey.**

26. **The SLM 2-10-0s of the Sfax-Gafsa company were its standard main-line engines for nearly 40 years.**

27. **This builder's photograph emphasises the classical lines of the Tunisian Pacifics. Nearly 300 metre-gauge steam locomotives worked in Tunisia over the years.**

28. **The metre-gauge privately-owned Sfax-Gafsa Railway operated a long line of distinguished types for its heavy phosphate traffic. The first effective engines were these two-cylinder compound 2-8-0s, supplied by SACM in 1904-05.**

28

World War II, the CFT built a line between Haïdra, on the Kalaa Djerda-Tebessa line, and Kasserine, on the Sousse-Henchir Souatir line. This connection reaches the altitude of 952 m – the highest of any Tunisian railway. Soon after it was completed in 1940 it proved useful, for it was the route used by the Sfax-Gafsa SLM 2-10-0s to Tebessa, where they were stored safely for the duration of the war. CFT 0-6-6-0s and 2-6-6-0s were also evacuated to Tebessa.

The Allied advance through Tunisia experienced the usual shortage of serviceable engines, so the Americans brought along their own metre-gauge power in the ubiquitous MacArthur-type WD 2-8-2s, which saw service in at least 12 countries during and after the war. Tunisia acquired 20 MacArthurs which presumably survived until the wholesale steam withdrawals in the 1950s.

MOROCCO

In 1859, following a dispute over the limits of the enclave of Ceuta, the Spanish waged a campaign against Morocco. General O'Donnell landed his troops about 25 km from Ceuta, on one of those sandy beaches where fertile river valleys break the monotony of cliffs along Morocco's Mediterranean coast. The nearby orange groves and town of Tetuan were tantalisingly visible from the sea, but the General must have wanted to conserve the energy of his men for more important things than transporting supplies, for he ordered an 11-km railway to be built up to Tetuan. It paid off and the Spanish won a decisive battle, though within two years British pressure forced them to leave.

Morocco was then in what the *Encyclopaedia Britannica* called a 'deplorable' state, with despotic and inefficient regional governments under hereditary sultans. It stated: 'Foreign commerce is hampered by vexatious prohibitions and restrictions, internal trade by the almost complete absence of roads and bridges, and by the generally lawless state of the country.' So it is small wonder that after the Tetuan line was abandoned, the Spanish railways in Morocco developed only in the 20th century.

For some time after the turn of the century it seemed that Morocco would be immune from the general scramble to control parts of Africa, the European powers having agreed at Algeciras in 1906 to give each other equal access to that territory. At the same time they undertook not to build standard-gauge railways there until a proposed line from Tangier to Fez had been completed. However, this did not prevent France and Spain, with an Algeciras mandate to guarantee order in Morocco, from building narrow-gauge lines and, eventually, dividing the country between them.

Iron ore in the Rif mountains proved a lode-stone and, in 1908, the Spanish Rif Mining Company built a metre-gauge line from the fortress port of Melilla to mines near Nador, while the North African Company built a 600 mm-gauge line which was later extended to mines at Cafra. Eventually, the Spanish protectorate government continued the metre-gauge lines to Tistutin for passengers and freight. At the same time, the Spanish authorities connected Ceuta and their new capital of Tetuan, also with a metre-gauge line. Of all these, the only line still in operation in the former Spanish zone is the first built.

When most of Morocco became a French protectorate in 1912, railway building had already begun. The first line was a short extension westward from Algeria to Oudja. Completed in October 1911, this was on 1055-mm gauge, as a concession to the Algeciras agreement, and had the distinction of becoming the first section of the modern standard-gauge system, being widened by 1916. It was worked by the Algerian Western.

A 600 mm-gauge military light railway, started from Casablanca in 1911, was the forerunner of the most extensive 600 mm network in Africa. Its peak route length of 1 700 km exceeded the systems of either German South West Africa or the Belgian Congo. These Moroccan lines were built under the direction of the French Resident-General, General Lyautey, to speed the distribution and provisioning of the Foreign Legion throughout what was then the French zone of Morocco.

Tracklaying was rapid, with minimal earthworks and fierce gradients, so that, with the exception of the line to Tangier (which was standard gauge from the start) what is practically the entire present-day network of the Moroccan Railways was laid down on 600-mm gauge by 1921.

Since then rail communication has traversed Morocco – and indeed the Barbary States, from Marrakesh in the west to Tunis in the east, a distance of 2 400 km. What a journey that would have been. There would have been a little physical discomfort – particularly on the bone-shaking narrow-gauge sections, and if one did not break one's journey for eating and sleeping. One would have started within sight of the snow-capped Atlas and ended at the city of Hannibal and St. Augustine, with relays of steam up front all the way; first perhaps, a Decauville Mallet, followed by a series of de Glehn compounds, then a Schneider 4-8-2, and a 4-cylinder Pacific on the last lap.

Although initially the Moroccan narrow-gauge lines were for military purposes, in 1916 they were opened to the public, passengers being conveyed in new Decauville saloon coaches, which W.J.K. Davies, in *Light Railways,* describes as 'extremely elegant'. One hopes that they were also comfortable, for travel on these lines was slow by any standards, with a maximum permitted speed of 25 km/h.

Since these were light railways it follows that almost the entire plant –

29

including pre-fabricated tracks and switches (consisting of 12-kg rail riveted to metal sleepers), locomotives, goods wagons and passenger coaches – was supplied by Decauville.

Apart from the standard Decauville 0-6-0Ts and 0-6-2Ts, the same firm supplied, in 1914, some interesting 0-6-6-0 Mallets with outside frames on the rear unit. They were simple expansion engines and used saturated steam with slide valves on all four cylinders driven by Walschaerts gear. They had a featherweight axleload of only 3 tonnes and consequently an incredibly low tractive effort for an articulated engine – only 4 545 kg (10 000 lb). After a decade they were sold out of service, one going to the Pithiviers-Toury tramway in northern France where, converted to a tank engine, it put in many years of work. Baldwin supplied some of their standard war-service 4-6-0Ts for light railways between 1915 and 1919. R.A.S. Abbott lists 19 Péchot-Bourdon type 0-4-0+0-4-0Ts built by Baldwin as having been shipped from France to Algeria in 1916. Though there were 600-mm lines in Algeria, it is possible that some or all of these Fairlies ended up in Morocco to assist the French military activity.

In 1921 SACM supplied 12 0-10-0Ts and these were followed in 1924-25 by the last new locomotives for the Moroccan 600-mm lines. There is some uncertainty about these, for Davies describes them as Decauville 2-6-6-0Ts, while E.D. Brant, in *Railways of North Africa* lists neither builder nor quantity, but gives their wheel arrangement as 2-6-6-2T.

In 1914, Lyautey, honouring the Algeciras agreement, negotiated with the Sultan for the construction by joint Franco-Spanish interests of a standard-gauge railway from Fez to Tangier, which had been since 1906 a free port in the Spanish zone.

The new company was referred to as the Tangier-Fez Railway Company (TF) and was known as such until after the end of steam in Morocco, though all trunk lines were operated as one system.

The outbreak of the World War I delayed construction and the first section of the TF, from Petitjean (now Sidi Kacem) via Meknes to Fez, was opened only in 1923. The TF took another four years to reach Tangier – an inexplicably long time in view of the easy terrain across the plain of Gharb, where the Romans had built a road along almost the same route.

Lyautey was awake to the limitations of the 600-mm lines as a national transport system. In 1920 the Moroccan Railway Company (Compagnie des Chemins de Fer du Maroc, or CFM) was formed to construct standard-gauge trunk routes to replace the 600-mm lines serving the principal towns. As in Algeria, the PLM had a substantial finger in the financial pie, this time in a consortium with the equally famous Paris-Orleans company (PO). The PO proved to be the dominant partner when CFM's motive power was modernized.

Unlike the TF, the new company did not let the grass grow under its sleepers and by 1923 standard-gauge trains were running between the new and old capitals, Rabat and Fez, utilizing the newly-completed section of the TF, from the junction at Petitjean to Fez. South of Rabat the standard gauge connected with Casablanca by 1925 and with Marrakesh by 1928.

A town which is dominated by a 12th-century minaret with a ramp instead of steps – making the top accessible on horseback – deserves something special in the way of railway stations. At least Lyautey and the town planners who helped him re-model Rabat evidently thought so. So as to 'respect the city of the sultan, with its high fawn walls, its mosques, its palaces, its harem, its royal gardens', and yet give the terminus from Tangier, Marrakesh, Casablanca and Fez a central position, and 'so that nothing shall be dirtied', they arranged for the new railway to approach the station through three tunnels, with the platforms below high embankments reached through a marble hall and flights of steps.

As new sections of standard-gauge route were opened, released 600-mm material was used to build new military branch lines. By 1921 a branch had been started up the Moulouya valley from the sheep centre of Guercif (on the Fez-Oujda line) to Outat el Hadj. Later it was extended to Midelt at the foot of the High Atlas, and many Foreign Legionaries used it to reach outlying mountains, oases and desert regions.

All other 600 mm branch lines were built between 1921 and 1931. These military lines must have been used extensively; during the 1925-26 Rif rebellion, for instance, Spain had an army of 100 000 in Morocco and France had 325 000.

The last major French campaign was in the Anti-Atlas and ended with the march on Tindouf in 1934. The military railways had served their purpose; the Moroccan people were united under a stable government; and many roads had been built. By 1935 all 600 mm branch lines were closed, the last to close being the first to have opened – the Guercif branch.

Only in 1934 was the last 600 mm track removed from the standard-gauge trunk route; for a long time the 'Imperial Main Line' included a very undignified 400 km stretch of 600 mm track between Fez and Oujda, involving a tremendous climb to negotiate the Atlantic-Mediterranean watershed. The

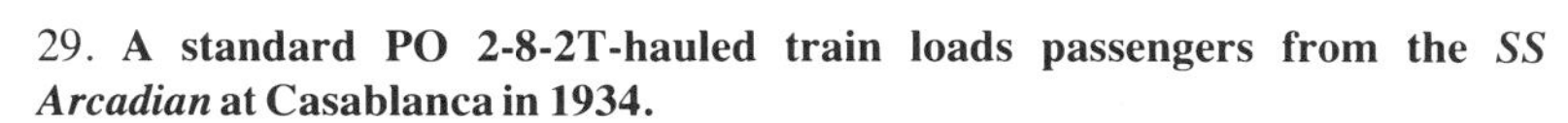

29. **A standard PO 2-8-2T-hauled train loads passengers from the *SS Arcadian* at Casablanca in 1934.**

30. **CFM 0-6-0 No 6, formerly of the PLM, is ready to haul excursion passengers to the docks after taking over the train from the electric locomotive at the left, in Casablanca, 1933.**

31. **In the Maghreb, Morocco had the smallest roster of steam locomotives, the standard-gauge stock totalling just over 100. The most successful main-line power on CFM was provided by these 2-8-2s, which were identical in design to the PO standard freight engines in mainland France during the inter-war period.**

30

31

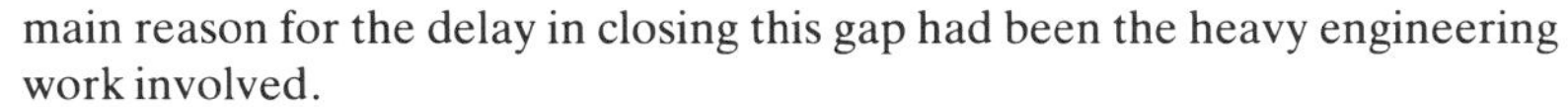

main reason for the delay in closing this gap had been the heavy engineering work involved.

Through the centuries, conquering armies had mustered at the cliff-hanging town of Taza to gather strength for the assault on Fez. They all had to go through the Taza Gap, a narrow passage between the Rif mountains in the north and the Atlas in the south; and the railway used the same gap, but even so found it tough going. Nowhere else in Africa at that time was such excellent railway civil engineering to be seen. The new line bridged many gorges and threaded many tunnels, with a total length of more than 10 km. The longest of these was the 2,5 km Touahar tunnel, which eliminated more than 400 m of rise and fall that the old 600 mm route had used to surmount the Touahar Pass. From 1934, passenger trains were able to offer Wagons-Lits sleeping and dining facilities all the way from Algiers to Casablanca.

Between the wars three important standard-gauge mineral branches which still exist were built. In 1923 the phosphate export conveyor from Khouribga to Casablanca was opened and later extended to the extensive deposits around Oued Zem. Another branch was built in 1936 from the Marrakesh line through phosphate-mining country to Safi, on the Atlantic seaboard. The third line reached from Oujda to Bou Arfa, with a spur to the coal mines at Hassiblal Jerada. Completed in 1931, its immediate purpose was to carry manganese, but it was also hoped that it would be the first leg of a Trans-Sahara railway; perhaps because of this it was diesel-operated from the start.

The standard gauge in Morocco was a late-comer to the railway family and these lines were never adorned by fashionable engines – there were too many older sisters. As each section of the standard-gauge line was completed, large numbers of hand-me-down 0-6-0s, 4-6-0s and 2-8-2Ts were sent both to the CFM and to the TF railways by their parent companies.

The CFM acquired 23 old PLM 0-6-0s, presumably used during construction work and later for shunting. Seventeen PO 5301 class 2-8-2Ts – built between 1914 and 1917 by Batignolles, SLM and North British – arrived for line service on the CFM during 1924-25. They had the rear section of the cab removed, as on a tender engine, and usually had one or two water-tank cars attached, Moroccan water plugs being further apart than in their native France. Twelve of the large PO two-cylinder simple 4-6-0s arrived for passenger work in 1927-28. The first and only new steam locomotives ordered for the Moroccan standard gauge were 12 Polish-built Mikados, to the same design as used on the PO, which were delivered to the CFM in 1932.

The TF began operations with five PLM 0-6-0s. As sections of the railway were opened, 21 2-8-2Ts and nine 4-6-0s were progressively acquired – both types being the same as used by the CFM. The last new steam locomotives to arrive in Morocco were eight American standard WD 2-8-0s, which the Allies imported during the war.

Steam cannot compete with electric traction fed by hydro-electricity, and as Moroccan water was of poor quality and coal was mined almost at the extreme eastern end of the railway system, the CFM engineers' decision to electrify was understandable. The PO had been a pioneer of main-line electrification in France so it is not surprising that in 1927 they initiated the first main-line electrification in North Africa.

The overhead catenary, energized at 3 000 V DC and fed by both hydro-electric and thermal power-stations, was extended rapidly after the first sec-

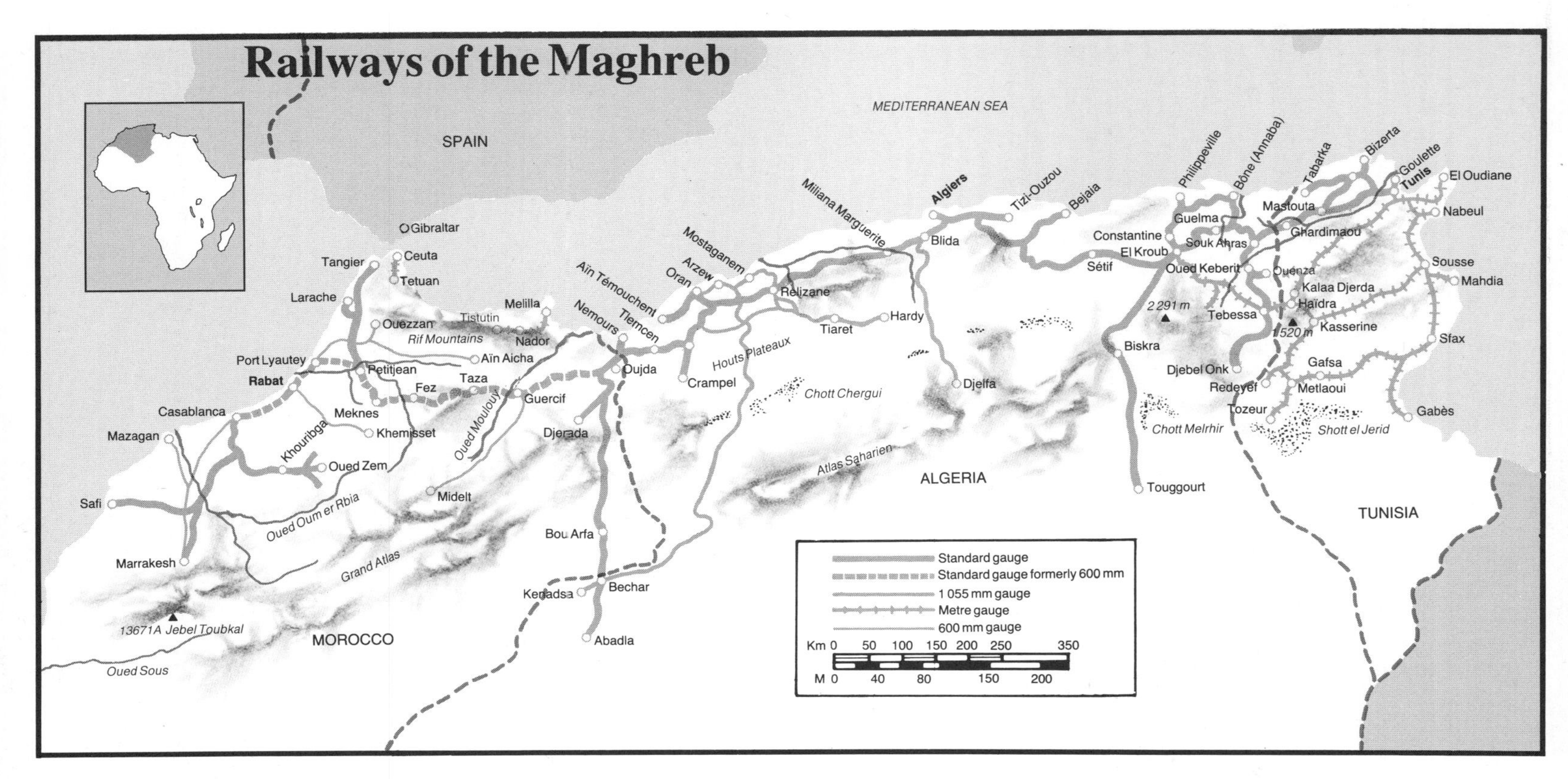

tion from Khouribga to Casablanca was switched on in 1927. By 1938 the main line from Marrakesh to Fez was under catenary, and steam was in full retreat, concentrated on the 360 km of the trunk route between Fez and Oujda, and between Petitjean and Tangier, with a few steam sheds retained in electrified areas for branches and shunting. World War II brought a temporary reprieve, but after the war diesel took over entirely. The last standard-gauge steam workings in Morocco ended in the late 1950s.

A tiny pocket of steam survived on two narrow gauges, truncated remains of the iron-ore railways around the old Spanish fortified port of Melilla. Early motive power here was provided by various small British-built tank engines, an Avonside 0-6-0T of 1900, two Kerr Stuart 4-6-0s and two Manning Wardle 0-6-0Ts. Some fine 2-8-2s, 2-6-2Ts and 4-8-4Ts were supplied by Henschel between 1920 and 1936, and some of these remained in operation until the 1970s, indeed, the very last steam in the Maghreb.

32

32. The metre-gauge railway from Ceuta to Tetuan was closed in 1959, though in 1933, when this 4-6-0 No. 3 (Alco 1913) was working at Ceuta, it was still a thriving concern.

33. Foreign legionaries stand, unconcerned, in front of one of the Decauville 0-6-6-0s at the 600 mm-gauge depot at Kenitra (Port Lyautey) during the early 1920s. By contrast, the train crew have a more soldierly stance beside the cab of their charge.

34. At least 27 of these Decauville Mallets were built for the 600 mm rails of Morocco. They were employed on the Atlantic-Mediterranean watershed section between Fez and Taza. It is estimated that close on 200 steam locomotives operated on the Moroccan Military railways.

34

33

35

36

37

LAST STEAM IN THE MAGHREB

35. Utilizing former military railway equipment, S.A. Minera Setolazar operated a 600 mm-gauge line in the former Spanish zone of Morocco and its steam locomotives were the last to operate in the Maghreb. The railway ran from the lead mines at Monte Afra in the Rif mountains to the Spanish port of Melilla. In 1966 an O&K 0-8-0, No. 1, 'Afra I', hauls an ore train through the streets of the town.

36. Two gauges were operated at Melilla. The Henschel 0-8-0T No. 12 in the foreground ran on 600 mm-gauge while the 2-6-2T, high on the ore-unloading terminal, is a metre-gauge locomotive No. 221 with iron ore from the Rif mines.

37. The metre-gauge lines belonged to the Compañia Espanola de Minas del Rif and employed two Henschel 2-6-2Ts built in 1920, which supplemented three earlier 0-6-0Ts. Even larger were three Henschel 4-8-4Ts, built in 1924, 1934 and 1936, one of which was noted in storage in 1966. Today the last remnant of the railway, which once ran almost 100 km inland, is operated by diesels.

38. Four fine 2-8-2s also graced the roster and were built by Borsig and Henschel in 1910 and 1921 respectively. This engine, No. 207, is the Henschel product.

39. A Henschel 0-8-0T, No. 12, 'Youksen II', shunts some ill-matched and decrepit wagons.

40. The utilitarian black of Afra I working an ore train through the streets of Melilla, contrasts sharply with the soft sandstone of the buildings which it passes.

38

39

40

41

2 EASTERN MEDITERRANEAN COUNTRIES

In Egypt's blistering Western Desert and the sand seas of Libya, where population is sparse and productivity scant, railways would be a pointless luxury. But in the fertile delta of the Nile and its mighty valley, agriculture has always flourished, supporting a teeming population and proving, early, the trackbuilders' old adage that railways should be roughly proportionate both to the number of a country's inhabitants and to their productivity.

It was in this rich delta, in 1852, that the first railway on the African continent was opened; it was from the network of rails which criss-crossed these silt-rich islands that two tentacles of metal track reached out – one southward up the Nile valley, the other to the west along the Mediterranean coast; and it was the head of this delta which would have formed the northern terminus of Cecil Rhodes' unrealized dream of a rail link from Cape to Cairo.

For the rest, Egypt remained as barren of railways as its western neighbour. Yet, though only slightly more than half the size of Libya, Egypt's 18-times larger population encouraged a spread of rail which today provides the country with a total route kilometrage of 4 740 km.

In sharp contrast, Libya, regarded as 'the poorest country in the world' until the comparatively recent discovery of rich oil deposits, has a rail map appropriate to this earlier status; a few scratchings along the coast – and even these have passed into history.

LIBYA

Believed to have evolved when someone laying metre-gauge track mistakenly assumed that gauge was measured from centre to centre instead of between each rail, the Italian 950 mm narrow-gauge system has an appropriately Latin independence, a flamboyant disregard for the norms of its

42

43

41. **A simple expansion, poppet-valved 0-4-4-0T Mallet, leaves Benghazi with a long mixed train, banked by a piston-valved version of the same class, on the Benghazi, Barce and Soluch Railway in 1945. These rare and interesting articulated refugees from Eritrea were built by Reggiane (Italy) in 1933 as Eritrean Railways 441.201 and 441.101 respectively.**
42. **The 0-8-0T was a common type on European narrow-gauge systems, and the Italian Class 401 naturally found itself working in Libya.**
43. **A 'streamlined' narrow-gauge tank engine. A Class 401 0-8-0T with armourplating for wartime protection – probably of psychological rather than practical value.**
44. **A local Libyan passenger train leaving Benghazi for Soluch, headed by a typical Italian narrow-gauge 2-6-0T No. 301.21, with an auxiliary tank car for desert working.**

44

46

47

48

45 (Previous page). **Operating 16 former Egyptian State Railways 2-4-2Ts, this was the last active steam-powered railway in Egypt when, in 1977, an Australian enthusiast, John Wilson, cycled a 20-km dirt track to Armant, on the west side of the Nile opposite Luxor, to record its working. When he returned two years later he found the entire standard-gauge railway abandoned and in its stead a narrow-gauge line powered by diesels.**

46. **Typical of the early Egyptian period is this old Stephenson 'Long Boiler' single-driver locomotive, for express work.**

47. **A turn of the century fumble towards increased power, the Trevithick outside-framed 4-4-0 was extended into Atlantic *Lady Cromer*.**

48. **Trevithick's outside-framed 0-6-0 goods engines survived almost to the end of main-line steam in Egypt. Here, one of these veterans, No. 702, rebuilt with piston valves, trundles an equally English-looking goods train near Giza in 1943.**

49. **Before the days of Rolls Royce limousines, vice-regal powered transport comprised this splendid steam carriage on rails.**

50. **A typical Egyptian passenger train of the twenties, headed by an elegant Trevithick 4-4-0.**

more conservative European neighbours. And, whether this explanation of the origins of this unusual gauge is apocryphal or not, this *mene se frega* attitude manifested itself in both the Libyan rail systems – the western, radiating from Tripoli, and the Cyrenaican to the east, based on Benghazi – for the lines were built during the Italian occupation.

The dearth of records for Libya's railways is in sad, but appropriate, parallel to the sparseness of the country's productive population; as elsewhere in North Africa, the laying of perways and tracks was governed by the availability of traffic. This was scant and both the three short lines centring on Tripoli and the two prongs extending from Benghazi had a total route kilometrage of not much more than 250 km each.

From Tripoli two short lines pushed along the coast, west to Zuara and eastward to Taguira, while a short distance along the Zuara line a branch led directly south to Garian. The hodge-podge nature of Libya's two systems, the Tripolitania and Cyrenaica Railways, is underlined by the intermittent and interchangeable deployment of its 33 locomotives which were switched from Tripolitania, to Cyrenaica and even Sicily with as gay an abandon as the railways appear to have been run.

The stock comprised 12 neat 0-8-0Ts built by the German Schwartzkopf works between 1908-12 and five typically Teutonic Hanomag 0-4-0Ts delivered in 1912-13. Italian locomotive works got a later, and smaller bite, of the stunted Libyan cherry, with seven of the smaller version of the 2-6-0Ts being built by Saronno and O.M. Milano, from 1912-14, and six of the larger, mainly by Saronno, from 1922-27.

Adding to the relative complexity of Libya's miniscule locomotive force were two Henschel 0-6-0Ts, built for the Austrian army in 1917, which found their way to the 600 mm-gauge line linking the local army camp with Tripoli.

Although 24 locomotives are known to have worked the two sections from Benghazi – one running northwards to Barce and the other south to Sluq – most were switched sporadically from Tripolitania. These were 0-4-0Ts, 2-6-0Ts and 0-8-0Ts. Four 0-4-4-0T Mallets were transferred from Eritrea and later confiscated by the British War Department; a similar fate touched three Garratt 2-8-2+2-8-2s built for Ethiopia in 1939 by Ansaldo, but diverted because of World War II.

THE EGYPTIAN STATE RAILWAY

Although the oldest railway system in Africa, the Egyptian State Railway (ESR) is, nevertheless, something of an anomaly. Other lines in Africa were developed to open up hitherto commercially virgin interiors; to extract easily and rapidly the riches of the hinterland; or to provide transport and matériel for frontline troops in the many minor wars which characterized the early years of colonial development. But the Egypt of the Pashas was relatively stable; the ever-flowing Nile provided transport for crops and produce; and there was little for the railway to serve in the way of heavy industry or minerals.

Then, as today, ESR mainly hauled passengers and agricultural produce, with the bulk of the traffic concentrated on the lines of the delta network.

There is an Arabian Nights quality to the locomotive history of the ESR. While there were none of the major absorptions, take-overs and changes of boundary that led to switches of gauge or rolling stock on many of the other African rail systems, the astounding range of non-standardized machinery in use during ESR's initial years proved a veritable works manager's nightmare. From 1852 to 1870 an incredible collection of 241 locomotives of 50 different classes was supplied by 16 builders from five countries.

These locomotives were essentially types common to the period, with 2-2-2s, 4-2-0s and 2-4-0s for express passenger work, 0-4-2s, 2-4-0s, 4-4-0s and 0-6-0s for slow passenger and goods work, a handful of 0-6-0STs for shunting and a few extraordinary 2-2-4T saloon locomotives for the conveyance of royalty. This miscellany was detailed by the noted locomotive his-

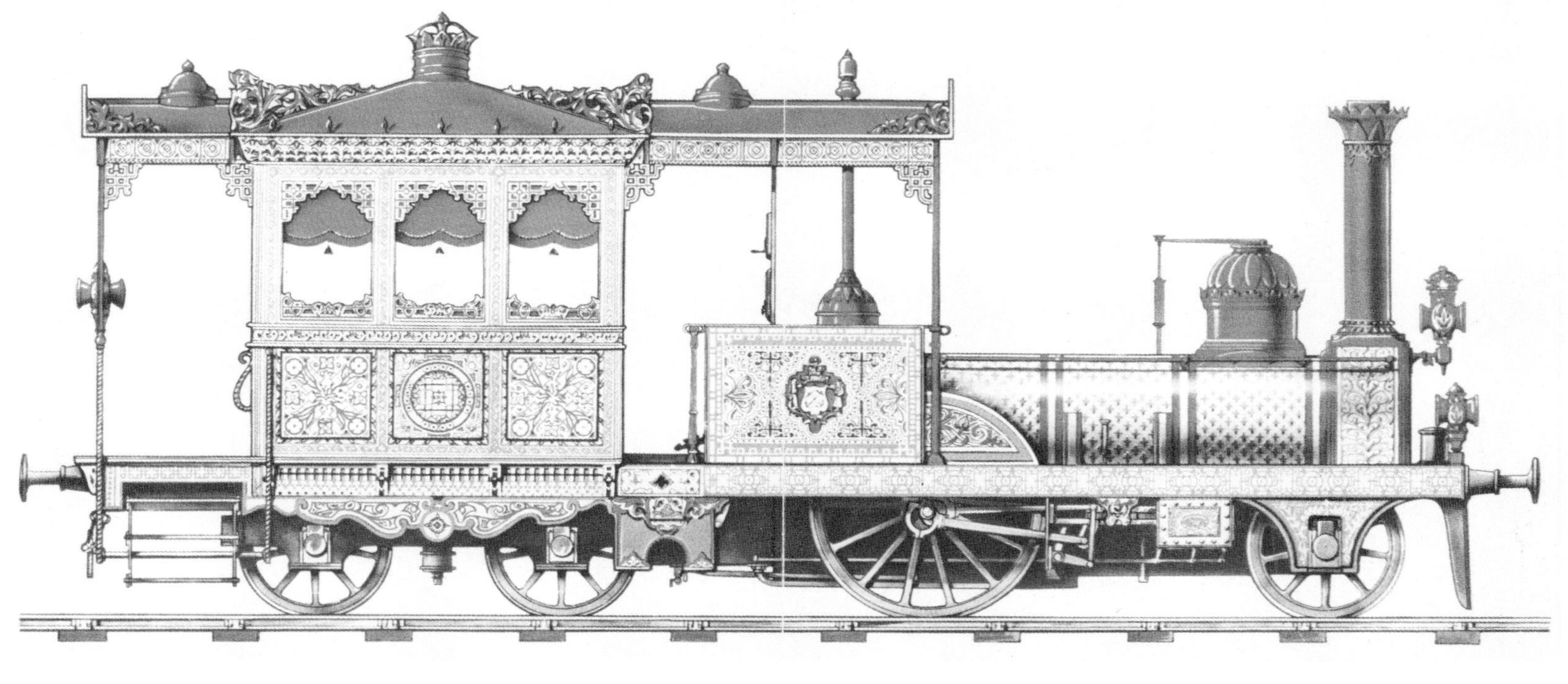

torian and former employee of ESR, E.L. Ahrons, in the *Locomotive Magazine* in a series of articles from 1903 to 1905 with supplementary material in 1913 and 1917-18.

The purchase of so heterogeneous a locomotive stock appears to have been at the whim of the Viceroy of Egypt, Said Pasha, who took a keen personal interest in his railway's motive power, but tended to pick what took his fancy in the same way, one assumes, that he would have chosen women for his harem. High utilization was unheard of and, in spite of the lack of standardization, engines were bought and the train service operated with remarkably few problems other than those endemic to a society geared to a less punctual and more leisurely way of life than that of Northern Europe.

Ahrons provides a delightfully understated record of the frustrations and problems stemming from such confrontations.

'The district superintendents in charge of the running sheds are as a rule Englishmen, and their position is no sinecure. Some of the native drivers are excellent men, but others are truly Oriental in their ideas of punctuality. It is not uncommon to have several goods trains ready to start and only half the necessary engine crews at hand for them. One man will send word at the last moment that he has a bad toe, and as he always takes the preliminary precaution of seeing a somewhat complaisant native doctor, he is in a position to smile and flourish a medical certificate in the face of the exasperated but powerless foreman. Another man, having married a wife, begs to be excused and it might be remembered that the Mohammedan engine driver can and does embark upon more matrimonial ventures than his British confrère. A third man, with a plethora of relations given to departing this life at more or less convenient intervals, buries one of them about once a month, produces a

50

51

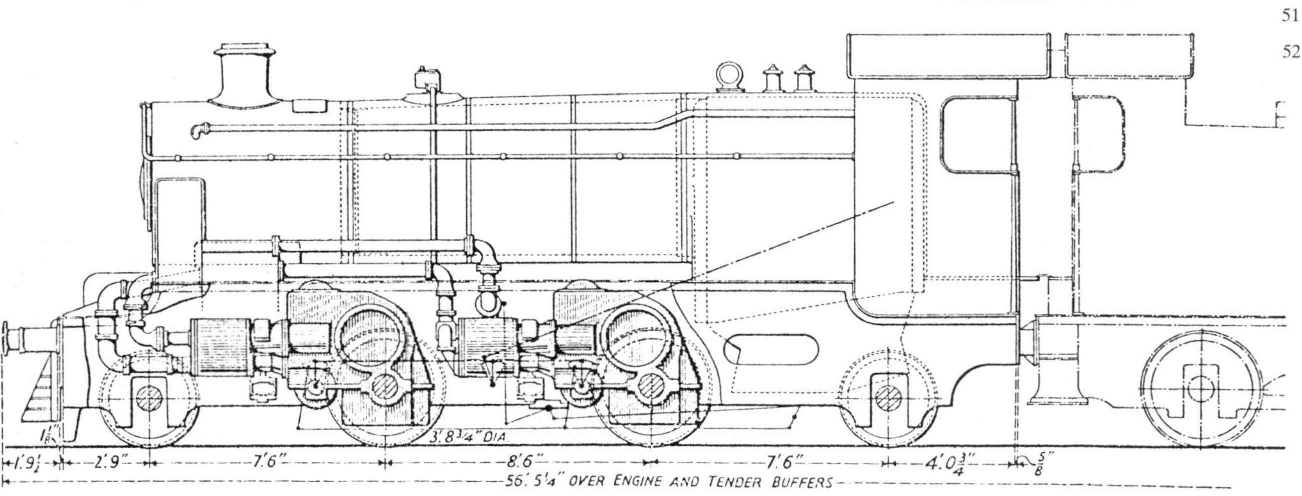

52

"certificate" to that effect and goes off smiling to the funeral. A fourth man has a religious excuse, and it is not always politic to interfere in such cases. Result, the six engines ready in steam to go out have perhaps two sets of men to go with them, and the foreman has to do the best he can to find a way out of the difficulty.

'The engine cleaner is another trying individual. When not engaged in saying his prayers in some remote corner of the running shed, the dim religious light of which renders it eminently adapted for temporary conversion into a mosque, he may perhaps bring his mind to a consideration of the question of wiping down locomotives from an academic point of view. He reasons that if engine No. 74 has to be cleaned by somebody, and if he, the cleaner, does not do it at once, it is evidently the will of heaven that it may be left until "to-morrow", unless of course somebody else does it in the meantime. Having duly thought this out, he possibly proceeds down the sidings for a quiet smoke, if he can do so without being discovered. Then he may be moved to have another look at No. 74, and should the foreman be in the vicinity, rubs a handrail very vigorously as being a part easily got at. As a result of a mental and physical shaking up administered by the foreman, our friend pays a certain amount of further attention to the engine until he again lapses into work in the abstract, half an hour after which he is probably to be found in the office asking for a rise of wages.'

If a plethora of motley locomotives characterized the first 18 years of ESR's existence, the following 18 years were marked by a dearth of new motive power. No engines were bought and it became an era of make-do and

53

51 & 52. **An interesting steam experiment was ESR's marriage of 'Sentinel' engine units to a conventional locomotive boiler, forming a 2-2-2-2 passenger engine. The drawing outlines mechanical details, and the photograph shows No. 278 on a passenger train at Cairo Main Station in 1946.**

53. **ESR No. 262 at Khatatba in 1942. Contemporary with the Sentinel 2-2-2-2 were a batch of 26 advanced lightweight 4-4-0s for similar local traffic. Superheated, and with Caprotti poppet valves, they fell early victims to diesel railcars. Twenty freight 2-6-0s were similarly equipped.**

mend. Again, a Viceregal whim was responsible for this state of affairs. Where Said Pasha had lavished enthusiasm and money – albeit haphazardly – on ESR, his successor, Ismail Pasha, showed no interest in the railway or its plight. The new Viceroy's attention was centred on building an extravagant new palace and filling it with Third Empire furniture – an obsession which absorbed not only most of his waking moments but also much of Egypt's finances, leaving no money to buy new locomotives or even to maintain those already in service.

The results of his financial prodigality were again recorded by Ahrons: 'The hard-worked heads of the locomotive department, mostly Englishmen, of whom Mr Garwood was the chief, and who carried upon their shoulders the burden of keeping the engines and rolling stock at work, had no sinecures. Let it be remembered that during the twelve years 1870 to 1882 hardly any new boilers were forthcoming to replace those which had been hard-worked for years. The Egyptian Government authorities of the period could not be made to understand why, given an engine or anything else on wheels, it should not be made to go, its state of repair being quite a minor consideration. Should it explode – "Never mind, get another engine; Allah wished that one to come to grief, and lo! see what has happened to it". The unlucky *employé*, or spectator – provided the latter was not a pasha – who may have been unfortunate enough to be hurried into eternity thereby, must needs have deserved his fate, or Providence would have reserved him for something else. Inshallah! everything therefore is for the best. But should such an unlucky contretemps have occurred as the transport to eternity per railway of some important pasha or bey, uncomfortable enquiries might be set on foot, which it was highly desirable to avoid.

'Those responsible for the locomotive stock did the best they could by reducing the boiler pressures from the original 120 or 140 lbs. to anything down to 80 lbs., that the boilers could safely stand. Afterwards there followed the events of 1883 – the landmark of modern Egyptian history – and the English railwaymen had to flee to Alexandria, where some of them fortified themselves inside Gabbari works.'

Britain's occupation of the Nile valley in 1882 led to a general re-organization of the Egyptian administration, but it was some time before improvements reached the ESR. The Khedive still teetered on the brink of bankruptcy and no funds were available to aid the faltering railways. An international loan of £9 million was negotiated in March 1885, but this was applied to staving off creditors rather than propping up ESR's perways or repairing its locomotives.

The first moves toward recovery came in the person of Mr F.H. Trevithick, who arrived from the Great Western Railway to head the locomotive department and 'generally straighten matters out'. He faced an uphill struggle despite the efforts, during the years of 'lean kine', of 'the little band of self-effacing Englishmen residing in the little-known native quarter of Boulac' and who, after his arrival, joined him in the hard task of 'rendering efficient the broken-down department for which they were responsible'. These included such railmen as Brown Bey, of the Boulac works, and Carlisle Bey, manager of the Gabbari works in Alexandria.

Trevithick reigned over the ESR locomotive department for 30 years and turned the run-down, ramshackle operation he inherited into a smoothly-functioning organization. The initial locomotives he introduced were essentially 'Great Western' in concept, with inside cylinders and double frames. More than familiarity or mere tradition probably lay behind this move. For, when the Great Western had converted from broad to standard gauge it had retained Brunel's original track laid on longitudinal sleepers – which gave a hard ride.

The British company found that in these conditions, double-framed locomotives provided the best results. It seems likely that Ismail Pasha's regime

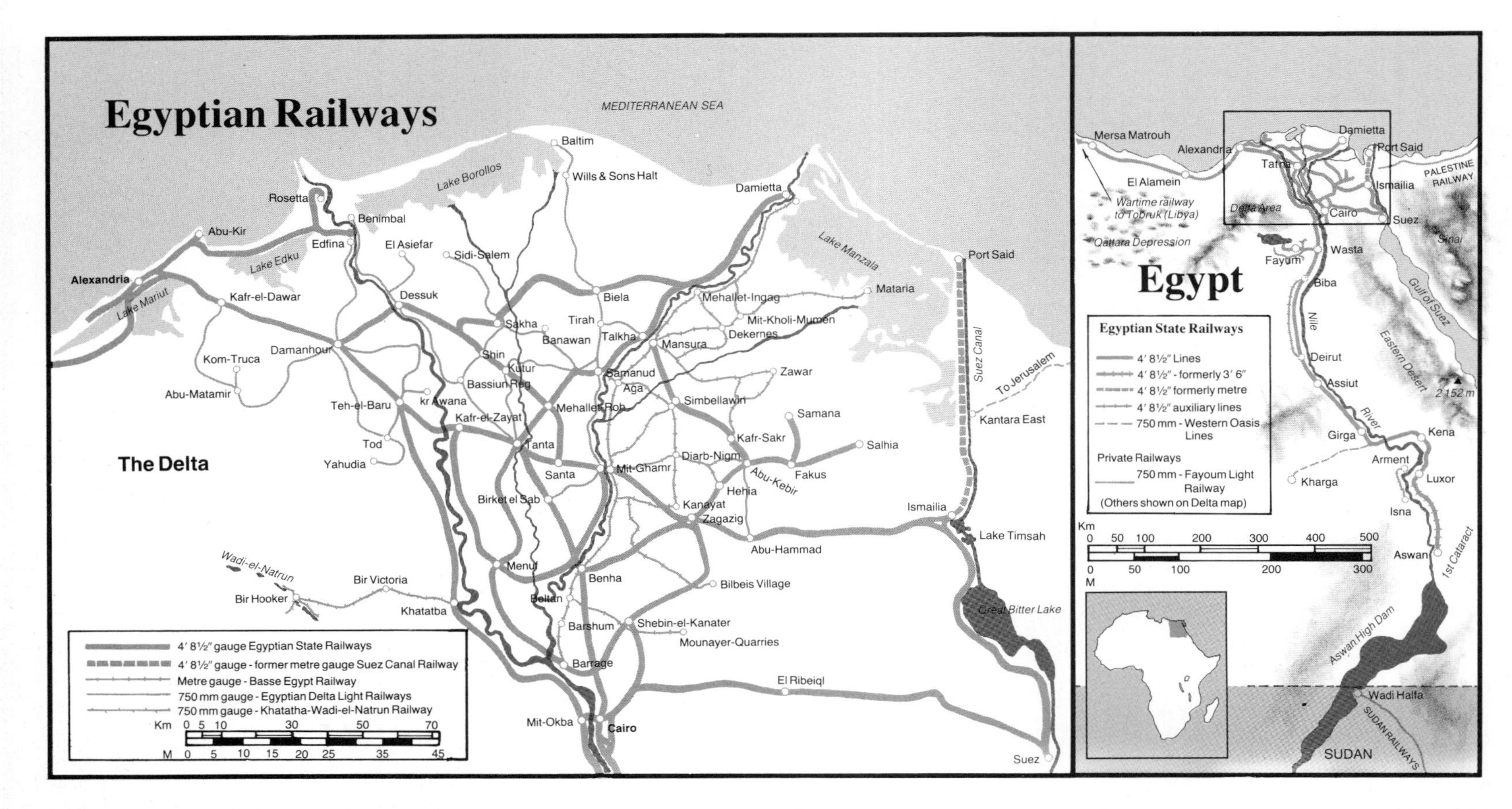

had allowed not only the locomotive stock but the trackwork to run down to such an extent that Trevithick found it useful to adopt double-framing as a palliative.

He set out to provide the ESR with a stud of conventional and reliable locomotives, without frills, to replace the previous models most of which were scrapped; a few, considered salvageable, were rebuilt with standard Trevithick boilers.

The locomotives of the Said Pasha era had been numbered from one upwards in order of arrival, which in view of the diversity of classes and types may have appeared to make sense, but Trevithick's orderly mind required the renumbering of stock into groups of digits for engines of similar types and duties.

Battling to bring some semblance of order to the ESR chaos, Trevithick would have had little time to contemplate the grandiose folly of the Isthmus of Suez Railway. Illustrated in the April 1884 issue of *Engineering,* this Brobdingnagian scheme proposed to convey ships bodily across the isthmus on five tracks of standard gauge, laid in parallel, with a total gauge between the outer tracks of 70 ft (21,34 m). The ships were to be hydraulically raised to rail level in a type of dry dock containing an enormous flat-wagon and then hauled across the isthmus by 20 locomotives, five abreast. The ingenuity of Victorian engineers knew no bounds, but it is perhaps as well that while the scheme was still on the drawing-boards it was superseded by the Suez Canal.

The canal, by increasing the French presence in Egypt, was also to exert a subtle, indirect influence on the development of ESR. As finance became available for Trevithick to buy new locomotive stock, Franco-Belge was among the first suppliers. For the lightest and fastest duties he ordered 25 single-drivers of the 2-2-2 type and these were built by Kitson and Franco-Belge from 1888 to 1894. They were the last 'singles' built for ESR and apart from the earlier ESR engines, were probably the only single-driver locomotives used in Africa. These, together with the fact that main lines were built to the European standard gauge of 4′ 8½″ (1 435 mm) gave the Egyptian rail system a character entirely different from other British-built railways in Africa. The ESR was, in effect, 'non-African' in concept, having more in common with the railways of India or Argentina. One traveller of the period even found it similar to Britain's railways, remarking: 'The very uniforms of the Egyptian stationmasters remind one of Waterloo and Victoria, the sole difference being the fez instead of the cap.'

A year after the delivery of the first standard 'singles' the first of 57 2-4-0s from Neilson and Franco-Belge were delivered and they continued to arrive until 1894. These double-framed 2-4-0s enjoyed so long an innings that a few were still in service during World War II. The 2-4-0 passenger engine was developed into a 4-4-0 version of which several essentially similar batches were built though varying in cylinder size and firebox dimensions.

The smaller versions – 41 locomotives in all – were built by Neilson and Henschel in 1901-02 and by North British in 1906; the larger were all Henschel-built in 1905-06, comprising 27 engines of normal design and two experimental locomotives, one with low-pressure boiler and large cylinders, the other with high pressure and small cylinders.

Although his standard locomotives were conventional, Trevithick had an unexpectedly inventive turn of mind and experimented extensively with superheaters and feed-water heating, producing some extraordinary, almost revolutionary, versions. One had a feed-water heating drum over the boiler and chimney over the cab, anticipating by several decades the design of the Franco-Crosti locomotives. Useful economies of up to 20% were claimed for Trevithick's designs, but they were taken no further than the experimental stage and, eventually, they were outclassed by more modern locomotives with conventional superheaters.

His standard goods engine was an outside framed 0-6-0 and, from 1892 to 1906, 180 of these were built to slightly different versions by a range of manufacturers. All Trevithick's locomotives had standardized boilers, cylinders and motion – not only within each class, but, wherever possible, between one class and another. His influence must have seemed a godsend to his hard-pressed workshop and running-shed staff.

When, in 1896, the traditional British builders were paralysed by strikes and continental factories swamped with orders – to keep traffic moving even several British railways had to accept locomotives designed and built in North America – Egypt sought the help of the Baldwin Locomotive Works to obtain 20 typically American 2-6-0 goods engines in 1898 and these were followed in 1900 by ten 4-4-0 passenger engines and ten 2-6-0T shunters.

Meanwhile, there was a need for heavier power to cope with 20th century traffic and, while standard engines were being churned out, several interesting experimental locomotives were bought. In the event, none of these was duplicated – but they did provide data on which decisions could be based. Four big engines were bought in 1900, two by Brooks and two by Dübs, to test the relative merits of British and American practice. Both Dübs engines were straightforward, typifying British design of the period, but diametrically the opposite of Trevithick's standards, having inside frames and outside cylinders.

No. 601 was a 4-4-2 passenger engine, and 701 a 2-8-0 for goods work. The Brooks engines were of comparable capacity, but whereas the 2-8-0, No. 700, was quite straightforward, 600 was designed to provide three alternatives in one, being built as a large-wheeled 4-4-2 for express work, but easily convertible to a 4-6-0 with either medium or small wheels for mixed traffic or goods. It is believed to have finished up as the goods 4-6-0 version. Apparently, none gave complete satisfaction – indeed, the eight-coupled types were possibly too powerful for the time, for nothing comparable was built for many years.

North British built two more experimental express locomotives in 1906, one an Atlantic, or 4-4-2, with inside cylinders and taper boiler; the other was similar but with three cylinders – an advanced concept for its day. Ten de Glehn compound 'Atlantics' of Nord design were supplied from France in 1905, and these, at least for a while, satisfied the fastest express train requirements. For slower expresses requiring more adhesion, ten inside-cylinder 4-6-0s were built by North British in 1908.

Goods traffic was handled by 20 large 0-6-0s with inside frames and cylinders of British style but, surprisingly, built by Henschel in 1907. These were considerably more powerful than the outside-framed type, but had a shorter life and disappeared before World War II.

Tank engines, at this time, were a rarity in Egypt, and, apart from a few odds and ends, were represented mainly by a class of 45 0-6-0ST shunters which doubled as suburban engines. In 1907, North British built ten inside-cylinder 4-4-2Ts for the Cairo suburban trains, and as these became rapidly outclassed, Henschel built 12 more powerful 2-6-2Ts in 1911, as well as four 0-8-0Ts for hump shunting at Alexandria. These Henschel engines, with outside cylinders and Walschaerts gear driving piston valves, were ESR's first really modern engines and set the standard for future development.

Modern Locomotives

For express passenger work, a capable 4-4-2 engine was evolved when, in 1913, the Berliner locomotive works built five initial engines. These, despite their German origins, were of essentially British appearance. With two outside cylinders, Walschaerts gear, piston valves and superheaters, they were straightforward and fully suited to their tasks – so much so that a further 75 similar engines, with slightly larger cylinders, were constructed between 1921-26 by British, German, French and American builders, the Baldwin batch being entirely standard including plate frames. These served as the main express engines for more than 20 years, until after World War II. In 1934-5, two were converted to 4-6-0s to improve adhesion; this pair served as prototypes for the numerous 4-6-0s introduced after World War II.

For goods work, the two-cylinder superheated 2-6-0 replaced as standard the outside framed 0-6-0 on main line work, with four of this new breed arriving quite fortuitously when a ship carrying them to the Baghdad railway was captured in the Mediterranean by British forces. The 2-6-0s forming part of the cargo were diverted to the ESR where they became useful additions to the ten similar engines built by Henschel and Franco-Belge in 1913-14. Thirty more of this design followed from Baldwin in 1921; like the Atlantics these were built to European standards with plate frames. Up to now, most engines had been built either with passenger or freight work in mind, but the 'mixed traffic' concept was fast gaining acceptance in Britain, with all four main-line companies building large numbers of medium power 2-6-0s with medium-size wheels suitable for a large range of duties.

Egypt also found this a useful proposition, so that 60 larger-wheeled mixed- traffic 2-6-0s were built by North British and Armstrong Whitworth in 1928, followed in 1931 by 20 similar engines from Borsig. The design was then revamped to give a lighter engine, although of similar power, with high boiler pressure and small cylinders. Thirty of these were built by North British in 1935-36, with domeless boilers, of which five had ACFI and five Heinl feedwater heaters. A further 20 had Caprotti poppet valves, forming a class of 50 basically similar locomotives. About this time, the 12 Henschel

54. **A local train, ready to start, hauled by an extraordinary combination of modern 2-6-4T and ancient 0-6-0. This scene is characteristic of the wide variety of power which comprised Egypt's all-time roster of nearly 1 450 locomotives – more than 60% of the total number of standard-gauge locomotives in Africa.**

55. **One of Egypt's 0-6-0 saddle tank shunters in a typical setting of lineside suburban Cairo.**

2-6-2Ts were converted to 2-6-0 tender engines, providing a total of 176 of these useful all-purpose machines on the ESR – second only to the outside-framed 0-6-0s in numbers.

The conversion of the earlier 2-6-2Ts to tender engines was not so much from a need for more suburban tank engines, but that they were being replaced in greater quantities on the main lines by more modern versions.

Three basic designs of modern tank engine were built, of which a small-wheeled 2-6-2T, mainly for short distance goods work, appeared in 20 Breda engines of 1923. These were followed by 60 similar engines from Britain and Belgium in 1927. The larger-wheeled version, for suburban work, manifested itself in 30 engines by Breda, built in 1924, and this again was developed into a large-wheeled 2-6-4T, having more water capacity, of which 40 were built by Breda and North British in 1929-30.

But the main requirement was not always for larger locomotives and, immediately before the war, ESR found it needed to replace various older engines on lightly-laid lines, and on light passenger work, where the more

56

modern and heavier types were clearly unnecessary. Accordingly, several interesting types were evolved, of which a 2-4-2T on a 12-tonne axle load – 15 of which were supplied by Bagnall in 1936, four from Skoda in 1939, and another eight after World War II by Bagnall – can best be described as a narrow gauge engine widened to standard, but with thoroughly modern detailing throughout. For light, fast passenger work on level lines, 26 modern 4-4-0 tender engines with Caprotti valve gear were built by North British in 1937, and these must surely have been the last new class of 4-4-0s introduced in quantity anywhere in the world, although 4-4-0s were subsequently built in the United States. The final light passenger type was unique among the world's locomotives, being an individual axle-drive 2-2-2-2 or, to describe this oddity in electrical locomotive designation, a 1-Bo-1.

At the time, there were several innovations to individual axle-drive steam locomotives, and experimental types were built in France and Germany, while others were projected in the USA. Only the Egyptian types, however, were in regular service. These saw the application of two standard Sentinel type power units to an otherwise normal tender engine in capacity similar to the light 4-4-0 previously described. The Sentinel high-speed geared engine had been evolved for use on heavy steam road lorries, and had then been developed into a light shunting unit for railways. The application of two or more of these high-efficiency units to drive a main-line steam locomotive was a promising development unfortunately curtailed by World War II and administered a post-war *coup de grâce* by high-pressure American diesel salesmanship. At the same time, some streamlined railcar units, also driven by Sentinel engines, and by high-pressure water tube boilers, were introduced onto the ESR, but it is uncertain whether they survived the war.

The Final Phase – Wartime and Post-War

Though Egypt had bought a substantial number of modern locomotives in the period between the two wars, ESR had never been called on to handle heavy goods traffic, and apart from the two experimental 2-8-0s of 1900, had not operated anything capable of handling extremely heavy freight. Thus the North African campaign of World War II found the railway unprepared. Trains carrying tanks, munitions and other matériel needed locomotives of substantially greater capacity than ESR's standard 2-6-0s and the War Department drafted large numbers of 2-8-0s to Egypt.

At the end of the war these were sold, no doubt at a favourable price, to ESR, and their greater capacities led to the rapid disposal of many older, and even newer but less powerful, engines. The precise details of these changes do not appear to have been recorded, but J.W.P. Rowledge has gathered such information as was available and from his series of booklets it seems that at least 80 of the Robinson 2-8-0s sent to the Middle East in 1941-42 became ESR property and that some of these were still operational until 1961.

Sixty-one of Stanier's 8F 2-8-0s used in Egypt were purchased by the ESR between 1942 and 1956, proving so popular that an Egyptian version was designed, using the chassis of the LMS 8F, surmounted by a smaller standard Egyptian boiler. While being standard, it may have facilitated workshop attention, but the boiler was unlikely to have steamed as well as the Stanier original. Engines of this 'Egyptian 8F' design were built by Vulcan Foundry and by MAVAG, of Hungary, about 1952, but the numbers are unknown.

With post-war freight haulage worked mainly by wartime 2-8-0s, ESR's

57

58

56. **A classic pre-World War II scene in Egypt, as Atlantic No. 42 whisks a light passenger train across a palm-studded plain between Ismalia and Port Said.**
57. **A fine engine portrait showing the handsome lines of an Egyptian Atlantic No. 56, named 'King Fouad I', on a bilingual English-Arabic nameplate.**
58. **Following the successful conversion of two Atlantics into 4-6-0s to improve adhesion, this type, in several versions, became the immediate post-1945 standard for heavy passenger work. A Transatlantic-featured 4-6-0 by Montreal was followed by typically Egyptian designs from Britain and France, the latter being pictured.**
59. **Egypt's final flowering of steam express locomotives were these handsome, French-built Pacifics, designed with a long, narrow firebox for oil firing. Arriving in 1955, they had a tragically short life in service.**

59

initial purchases were for heavy passenger engines, needed to replace the Atlantics of the previous generation and to cater for the additional heavy traffic. This was done with some vigour and following the satisfactory adhesion characteristics of the Atlantics rebuilt to 4-6-0, this latter wheel arrangement was made standard. The first was a batch built by the Montreal Locomotive Co., in 1948, to thoroughly North American standards. These were followed in 1949 by 80 engines from North British, with generally similar basic dimensions, but the advantage of plate-frame construction to provide a long narrow firebox specially adapted to oil firing, using the low grade 'Mazout' oil normally supplied to the ESR. A further ten locos, with even longer fireboxes were supplied by Franco-Belge in 1951-52 in an attempt to provide more horsepower, and ESR's final steam locomotives were 20 Pacifics built by SACM in 1955. These were almost built as Hudsons, but the original specification, calling for a two-hour schedule for the 150 km from Cairo to Alexandria with 550 tonne trains, was eased to suit a load of 500 tonnes, with the result that a 4-6-2 sufficed. The design of these engines is unusual for a Pacific as, being specified entirely for oil fuel, a long narrow firebox with combustion chamber was fitted between plate frames. It is not known exactly how long these fine engines and other more modern classes lasted in service, as for several years ESR refused to admit that it operated steam locomotives, in the mistaken belief that by giving the impression their locomotives were 'all diesel' they were 'up to date' and 'modern'.

The Suez Canal Railway

Between Port Said and Ismailia, the Suez Canal Co. ran a metre-gauge railway, opened by the Khedive in 1893. The stations were numbered, to coincide with the canal's signal stations and the line carried mainly mail and passenger traffic. Only eight French-built locomotives were in use, all with outside frames and cylinders. The first two were 0-6-0s by Corpet Louvet, used for goods and ballast work, and the remainder were passenger engines by SACM, four 2-4-0s of 1893 and two 4-4-0s of 1896 being numbered V1 to V6, 'V' standing for *voyageurs,* or passengers. The line was converted to standard gauge about 1907, and became part of the ESR, while most of the locomotives passed to the Delta Light Railways, where they were presumably changed to 750 mm gauge, a relatively simple conversion with an outside framed design.

Luxor Assuan Railway

This was the southward extension of the ESR main line up the Nile Valley, and presumably was built to 3′ 6″ gauge to connect, eventually, with the Sudan system, although this did not materialize. It was opened in 1898, and the first passenger engines were 4-4-0s with inside cylinders and outside frames, similar to the Ajmer design of the BBCI in India. The majority of locomotives used were Baldwin 2-6-0 tender engines, and in 1912, the same firm built two 0-8-0Ts for heavy shunting. Later, some 4-6-0 passenger engines were built by Maffei, and eventually the line was converted to standard gauge.

Egyptian Delta Light Railways

This extensive system eventually comprised about 1 000 km of 750 mm-gauge track, acting as feeders to the main lines in the Nile Delta, and serving the agricultural requirements of this fertile area. The first locomotives were 35 very neat 4-4-0Ts built by Bagnall in 1898, while from 1898 to 1900 Krauss built 18 2-4-0Ts and 16 0-6-2Ts, indicating a rapid initial construction of the various lines, not all of which interconnected. Nasmyth Wilson and North British later built 30 0-6-4Ts, between 1903-07. Further motive power was

60

acquired steadily, for 129 locos are listed in 1930, and later about 50 Sentinel locomotives were bought. In the 1930s, some diesels were added. The whole system is believed closed as road competition has made it uneconomic.

Other Light Railways

Several other light railways existed in Egypt, of which the principal was the C. de F. du Basse Egypt, a metre-gauge version of the Delta system, with about 22 locomotives running on 257 km of track. This Brussels-based company probably used light Belgian-built locomotives. Also Brussels-based, the Fayoum Light Railways, of 750 mm gauge, had 17 locos on 168 km of track; while outside the Delta area, the Egyptian Soda and Salt Co. ran a 56 km line out into the desert at Wadi el Natrun.

Altogether, Egypt, in the inter-war period, must have been a fascinating place for the light railway enthusiast; but, sadly, little detail of these lines was recorded before they were swamped by road transport.

60. **The Corporation of Western Egypt ran an extensive 750 mm-gauge system into the western oases, motive power including this neat Nasmyth Wilson 0-4-2 of typically British colonial outline.**

61. **The maid-of-all-work in Egypt was the medium-sized 2-6-0. Built in several variations to suit goods or mixed traffic, the later examples had poppet valves. This small-wheeled, piston-valve version is one of the earlier engines.**

62. **During World War II, large numbers of British 2-8-0 goods engines were drafted into Egypt to cope with the heavy traffic. Many were Stanier engines of LMS design, such as No. 432 shown here, looking slightly self-conscious wearing cowcatcher, oil burner, and auxiliary water tank at El Daba on the Western Desert Railway in 1946.**

63. **The Delta Light Railways made extensive use of Sentinel four-wheeled geared steam locomotives. This engine, No. 210, heads a train from Benha to Barrage out of Barshdum station. Though the carriages are French-looking, the setting is a typical Egyptian background of palm trees, mosque, and suitably-garbed local inhabitants.**

61 62

63

EAST AFRICA

3 DESERT LANDS OF THE NORTH

Of the four diverse countries making up the desert lands of the north, only three today operate railways. Two of these, Ethiopia and Djibouti, are linked by virtue of Djibouti, a former French territory, having played host to Ethiopia as an outlet to the sea. A mere 100 km of railway exists in Djibouti and, as it is operated by Ethiopia, can hardly be said to have a character of its own. Sudan's railway system is the largest and most developed in the region; while Somaliland, occupying most of the coast of the Horn of Africa, once had railways, but today is one of a growing number of countries where rail transport has become history.

European colonization has left its uneasy mark and the heroic feats of men such as Gordon and Kitchener in the Sudan are hardly remembered today. Even though no Italian generals left their names in the sands of the Horn of Africa, the Italian occupation of the area created masterpieces of railway construction and one of the world's most impressive railways. The area has lately suffered greatly from the depredations of war, but in this part of Africa – where both climate and temperament tend to the extremes – it is perhaps fitting that the railway, born in fire, should end in fire.

SUDAN

The Mahdi's hordes swept north through the Nubian Desert, laying waste to some 160 km of railway and equipment before they halted at the border with Egypt. And so the first major railway project in the Sudan was destroyed – barely ten years after its inauguration in 1875. Khartoum and General Gordon's garrison had fallen in the *Jehad* and it would be another decade before the British regained control and railway building could recommence.

This second railway was strictly military in purpose. Between 1896-98, General Kitchener's forces pushed south, constructing first a 315-km line along the Nile to Kerma, at the Third Cataract, then a 600-km line through the Nubian Desert to Atbara, and finally a 320-km extension to Khartoum. With the defeat of the Mahdi's followers, the area was pacified – leaving Sudan with a unique railway system which, though valuable as a line of military communication, was quite unsuitable as a main artery of trade.

Starting in the north below the Second Cataract at Wadi Halfa, the line was 1 600 km from the mouth of the Nile and nearly 500 km from the Red Sea. Traffic moving to the Sudan from Egypt involved trans-shipment from standard-gauge rails to river transport, then back to 1 067-mm gauge rails in Sudan – an expensive and time-consuming process. Clearly, a connection from the Nile Valley to the Red Sea was needed and the old caravan road between Suakin, on the coast, and Berber, on the Nile, offered a ready-made route. In 1905 this line was completed and as such it marked the beginning of non-military rail development in the Sudan – the 'third phase', described by Leo Weinthal in *The Story of the Cape to Cairo Railway:*

'It may be said that the story of railway building in the Sudan falls into three main periods. *First,* the period of failure, when great schemes were conceived and initiated, only to be abandoned either for financial or political reasons. Secondly, the period of conquest, when railway building and military operations were mutually dependent and mutually essential; and thirdly, the period of development which followed upon the re-establishment of order and settled government.'

Many of Africa's railways fall into a similar pattern and those of Sudan's eastern neighbours – Ethiopia and Eritrea – are no exception. All three are covered by large tracts of desert, and the romantic tales of military conflict which have filtered down over the years have created illusory concepts about the harshness of the region. Substantial areas are fertile, though not particularly lush.

Nevertheless, Kitchener's epic crossing of the Nubian Desert provided a topic of conversation, particularly amongst practical-minded people of the day. Many questioned the logic which led to the building of railways across trackless deserts – seemingly from one mirage to another.

Mad dogs and Englishmen . . .? At least these 'Englishmen' left the native Sudanese with the foundations of a modern railway system, though transportation was still largely on the backs of camels or on the waters of the Nile. As

64. A busy scene at Sennar, where 0-6-0T No. 13 and 2-8-2 No. 225 shunt, while in the background a 220 Class Pacific prepares to leave with a passenger train.

65. In August 1976 Nils Huxtable rode from Babanousa to Malwal, detraining to photograph the light 2-8-2 No. 322 starting its load – which included more passengers on the carriage roofs than many a train has inside.

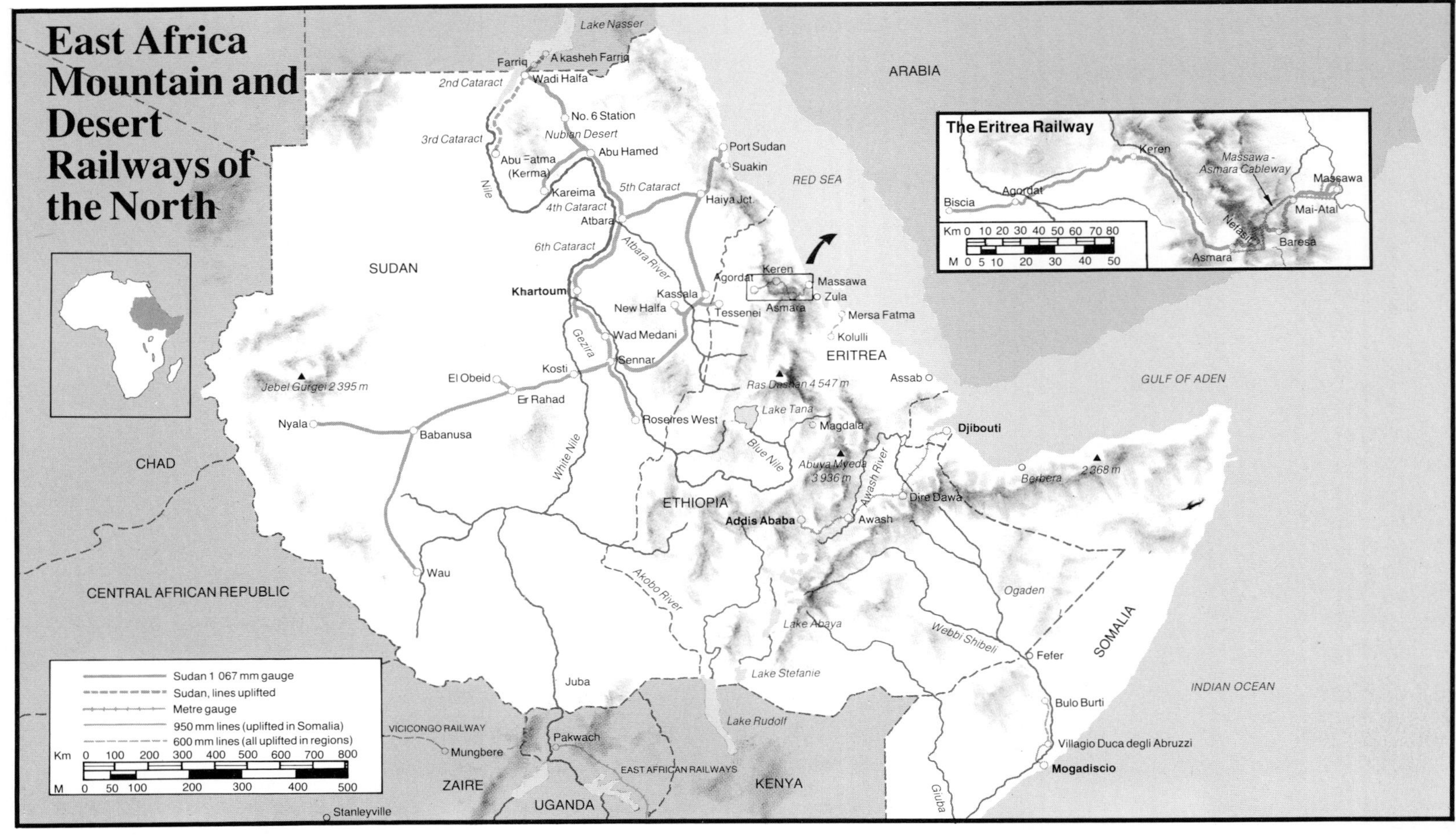

in Egypt, the waters of the great river were essential to agriculture. Sudan was particularly fortunate in that the Nile forks at Khartoum. The White Nile flows from numerous tributaries starting in the Congo and Uganda, through the south-west corner of the country, while the Blue Nile drains the Ethiopian highlands; the two form a rough triangle of extremely fertile land which only needed proper irrigation to produce agricultural abundance.

Development was the catch-word and the opening of the Suakin line from the Red Sea stimulated later expansion in southern Sudan. In earlier days this line from Suakin was part of yet another military venture and was notable for the fact that the builders laid it with a gauge of 1 435 mm. General Sir Gerald Graham built the short, 30-km line inland from the coast early in the campaign to relieve Khartoum. But it was too late and, after continual harassment by Dervishes, the line was dismantled and the locomotives returned to England in 1885. However, it was the beginning of the Red Sea connection and when construction was resumed early in the 1900s, the old standard-gauge formation was utilized, though soon afterwards a more favourable site for a harbour was discovered 45 km to the north. In 1906 a new line was built to this site, later known as Port Sudan, and eventually the original line to Suakin was abandoned.

Rail development in the south began in earnest when, in 1909, the main line was extended along the Blue Nile to Sennar, and in 1912 was continued westwards and across the White Nile to El Obeid. Soon before the outbreak of World War I, a short 10-km branch was built from Sennar to Makwa to carry material for the Sennar Dam on the Blue Nile, after which construction projects ceased for nearly ten years.

As if to make up for this static period, in the early 1920s new lines were proposed rapidly and the most ambitious idea actually came to fruition. This was an 850-km line from Haiya Junction on the Port Sudan-Atbara line, south through Kassala into the cotton-growing Gash area, on across the Atbara River on a seven-span 1 663-m bridge, and finally running across the Sennar Dam wall to Sennar. Begun at the north end in 1920, the line was completed by 1929. Passing close to the borders of Eritrea and Ethiopia, it was to have considerable strategic value several years later with the outbreak of World War II. In 1941 the Allies planned to invade Italian-controlled Eritrea and constructed a branch, extending from Malawiya into Eritrea at Tessenei. A mere 160-km gap separated the Sudan railway system from the Eritrean at Agordat, but the military chose to use road transport instead and the rail connection was never made.

A final burst of railway activity began in 1952 when a new line was opened south from Sennar along the Blue Nile for 220 km to Roseres, site of yet another Nile dam. Then, from 1956 to 1961, some 1 000 km of new line were opened in the south-west – extending from Er Rahad and forking, one prong running west towards the Chad border and the other south towards the Zaire and Uganda borders, terminating at Wau. Since then there have been numerous proposals to extend this line to Juba on the White Nile – just 80 km short of the Ugandan border, where a connection could be made with EAR's Pakwach branch, which extends to within 125 km of the border. Although a mere 500 km now separates the two railways, this link remains as much of a dream as the Cape-to-Cairo railway it would facilitate.

Much of the Sudan is flat and relatively featureless, and the character of many of its locomotives seems to reflect this landscape. With the exception of the Port Sudan line, most lines are easily graded and do not require large locomotives. Information on the original Wadi Haifa, and even the Suakin military railway, is obscure, more being recorded of the locomotives used during the 1896-98 campaign.

The first of these was a compact and powerful 2-6-2T named 'Endeavour'. Seven additional engines of this type followed, being delivered in Alexandria and taken up the Nile by boat and rail to Wadi Haifa. Engine water supplies were a major problem as the 'Soudan Expedition Railway' pushed south into the Nubian Desert. It was calculated that if no water was found along the nearly 400 km of line, each engine would have to pull 16 water tanks of 13 500 ℓ (3 000 gal.) across the desert – merely for its own use. Fortunately, water was found at mile post 77 and also at 126, though this did not stop the authorities from experimenting with three 2-6-2Ts which were fitted with a condensing vehicle immediately behind the locomotive. Details of the performance of these engines have not come to light – presumably lost in the

66. **Built for the Soudan Expedition Railway, this engine was originally pictured in the October 23, 1896 issue of *The Engineer*.**
67. **The origin of Sudan's early American locomotives is obscure. There were both 2-6-0s and 4-4-0s, and one of the latter is shown 'conveying the Sirdar to the Front' – obviously taken during the time of Kitchener.**

sands of time or the Nubian Desert. Interestingly, stations along this railway were numbered – not named – perhaps a reflection of the lack of human habitation in this barren area.

After 1900, when the railway came under civilian management, locomotives that would have looked at home in the South African Karoo were introduced. These included two classes of 4-6-0s, the last of which were constructed in 1907. Among Sudan's earliest locomotives were several American 4-4-0s and 2-6-0s constructed in 1896 and acquired possibly as a result of the British locomotive builders' strikes which necessitated the purchase of 'Yankee' products.

Following delivery of the 4-6-0s, the railway reversed the usual trend towards locomotives with additional coupled-wheels by ordering four very smart 4-4-2s with 1 588 mm (5′ 2″) coupled-wheels, slide valves, belpare fireboxes and generous sized cabs. Seen on trains of ivory-painted wooden coaches equipped with permanent louvres extending the full length of the sides, a more typical Sudanese railway sight is hard to imagine. Unfortunately, these fine engines were retired soon after World War II, and the latest Sudanese coaches – though still painted ivory – are now all steel.

Locomotives which were to have a lasting impact first appeared in 1911, these being the 120 Class 4-6-2s. Built by North British, these lightweight engines had distinctive low slung boilers and relatively large coupled-wheels measuring 1 372 mm (4′ 6″). They had a tractive effort of 9 716 kg (21 420 lbs) and an axle-load of 12 tonnes – suitable for the many miles of 25 kg (50 lb) rail in use at the time. The 15 of these general service engines which were built were the forerunners of Sudan's most numerous locomotive type, the 220 Class 4-6-2s. First built in 1927, a total of 51 engines were added to the roster by 1947 and the design was utilized by the British War Department as a 'standard', examples later being constructed for the Nyasaland and Trans-Zambesia Railways, as well as the Western Australian Railways.

In later years, and until the mid-1970s, most of these engines remained in service, concentrated mainly on the southern branches which had old 25 kg (50 lb) rail, removed from the north after the Port Sudan-Khartoum-El Obeid main line had been relaid with 37,5 kg (75 lb) material.

For heavy freight service on the Port Sudan line, the 150 Class 2-8-2s were introduced in 1920, ten being built by North British. Nine more engines of the same type were constructed by Robert Stephenson in 1926-29 and until the advent of the 4-8-2s in the 1950s they were the heaviest and most powerful non-articulated locomotives in the Sudan. Since their sphere of operation was limited, a lighter version was designed, and the first of this new 180 class appeared in 1923. A later version, the 310 Class numbering 19 locomotives, was constructed in 1952. Altogether 80 locomotives of three classes (180, 220, 310) all utilized the same boiler and other standard interchangeable parts.

An unusual type which appeared in the mid-1920s was the 200 Class. Thirteen of these non-standard Prairies were built for passenger service, supplementing the earlier Atlantics, and having in common with them 1 588 mm (5′ 2″) coupled wheels as well as a 14 545 kg (16 tonne) axle load, which restricted them to routes laid with 37,5 kg rail.

Purely shunting classes were few and all were six-coupled tank engines.

68. **Though these 0-6-0Ts were small, with a mass of only 40 tonnes, they proved popular as shunting engines and the type continued to be produced from 1927 to 1951. Some may still be in harness today.**
69. **A modern, but decrepit-looking steam engine is passed by an age-old method of transport. And long after the last steamer has laid its ashes to rest in the Sudan, animal power will still be there.**
70. **Power – in the Sudanese mould. A large post-war 4-8-2 sits in readiness at Kosti, crossing place of the White Nile on the line from Sennar to El Cheid.**

68

69

No
514

71

72

73

74

First came the 21 Class 0-6-2Ts, six being built by Manning Wardle in 1923 and 1929. A lighter class of 0-6-0Ts, the 7th Class, constructed by Hunslet appeared in 1927 and repeat orders for these small engines continued to 1951 – more than 30 being built. The only class of tank engines not built for shunting services in this period was the 31 Class 2-6-4Ts, four of which were constructed in 1931 by Kitson. However, when last seen in service during the mid-1970s they, too, had been relegated to this humble service.

Perhaps the most remarkable steam locomotives to operate in the Sudan were articulated – a single class of Garratts, which introduced for the first time the 4-6-4 + 4-6-4 wheel arrangement. Even though they had 1 448 mm (4′ 9″) coupled wheels, they were designed for freight and the 'passenger' wheel arrangement was introduced not for speed, but water capacity. It was planned to use these engines on the hilly Port Sudan-Haiya Junction-Atbara section where heavy Mikados were in use. On this mainline section the 2-8-2s could operate between water points without needing extra water tanks and it was planned that the Garratts should do the same, thus hauling the maximum nett load. The second idea was to operate these engines on caboose workings with through trains from Atbara to Khartoum on 37,5 kg rail, then beyond to Wad Medani on 25 kg rail. This would obviate the need for an engine change at Khartoum and involved a 1 000 km round trip. The engines hauled 1 600-tonne trains in this service and at the time of construction were the world's most powerful locomotives operating on 25 kg rail. With a tractive effort of 19 740 kg (43 250 lbs) they were the most powerful locomotives ever to operate in Sudan, but their use here was short, even though much fanfare was made of their design. In 1949 they were sold to the Rhodesian Railways who, with a large fleet of Garratts, seemed more favourably disposed towards these engines, although they resold them in 1964 to the Moçambique State Railways (CFM) where it is believed they still operate. One suspects that there was nothing inherently wrong with the design – although for the Sudan a simple, rugged, non-articulated engine seemed better suited to local conditions – for the railway had to contend with unskilled labour in the workshops and drifting sand out on the line.

World War II created a severe strain on the Sudan system and, apart from the delivery of more 220 Class Pacifics, the South African Railways supplied 16 second-hand locomotives, the oldest of which were built several years before the creation of the 'modern' Sudan system. These were 6th Class engines, built in England for the Cape Government Railways from 1893 to 1898, and similar in size and design to Sudan's own early 4-6-0s. They were coal-fired – as were most other contemporary Sudanese locomotives which burned fuel mined mainly in Natal, South Africa. Old and war-weary, these former SAR engines were all retired by 1950.

The first post-war locomotives were merely repeat orders of pre-war types: more class 220 and class 310 which was identical to the class 180 – for road service; for shunting, the 40 Class – identical to the earlier 7th Class – was obtained. With a rapid increase in traffic, and a programme to extend 37,5 kg

75

rails beyond the Port Sudan-Khartoum section to Sennar and El Obeid, larger and heavier non-articulated engines could be used. An order was placed for a heavy 4-8-2 type to be constructed by North British – looking suspiciously like a South African Railways 15CA, of which North British had built 47 during the late 1920s. In fact, the Sudan engines – 500 Class – were considerably smaller, weighing 88 tonnes, compared to 106 tonnes for the 15CAs. Nevertheless, they were the largest non-articulated engines to run in Sudan, and 42 were built in 1954. At first used between Port Sudan and Khartoum, they moved south on freshly-laid heavy rail in the early 1960s as dieselization progressed. In the early 1970s they were still intensively used between Khartoum and El Obeid.

Sudan was one of Africa's earliest diesel-users, obtaining its first models in 1936, but these were only small shunters used at Port Sudan, and up to 1959 all road services were handled by steam locomotives. When the first big diesels arrived, it was natural for them to operate on the more arid sections and the Port Sudan line was first to dieselize. By 1962 all through services between Port Sudan and Khartoum were diesel and within three years the Haiya-Sennar line had followed suit. More and more diesels came into services in the late 1960s and early 1970s but – here, as in many developing countries – old ones were breaking down almost as fast as new engines arrived. Severe foreign exchange problems reduced diesel locomotive availability through parts shortages to such a point that trains had to be cancelled, while on nearby tracks dozens of steamers stood silent – also waiting for repairs, some very minor, which could not be carried out as most shop space had already gone over to diesel repairs.

Operating performance on any railway is the most important criterion of its effectiveness, and statistics indicate that in recent years Sudan has not done well, diesels notwithstanding. Considering that major rail route expansion has taken place in the last 40 years – increasing the route kilometrage from 3 191 in 1938 to 4 734 in 1977; and that locomotives have increased in number from 141 (135 steam) in 1938 to 206 (180 steam) in 1961; and 359 (127 steam, 232 diesel) in 1978 – the following traffic figures prove revealing. In 1938, freight tonnes amounted to 681 000 and passenger journeys to 1 089 000. By 1961, freight had increased four-fold to 2 500 000 tonnes while passenger traffic increased three-fold to more than 3 million. However, by 1978 freight traffic stood at only 2 004 000 tonnes (after reaching a peak of more than 3 million in the early 1970s) and passenger traffic remained static at 3 million.

Steam which was scheduled to be retired was retained in the well-watered veld of the south and, even as late as early 1979, could be seen on passenger and freight trains. All this has changed, for in 1979 more 'growlers' were obtained and the railway was able to announce 'complete dieselization' – although 'a few' steam engines were in occasional shunting service. In view of experience elsewhere, it remains to be seen when the last steam engine in the Sudan will finally drop its fire.

71. **Graceful lines of a Cape-gauge Atlantic, named 'Sirdar', after Kitchener of Khartoum.**
72. **Sudan's first 4-6-2s were these dainty engines, smaller even than similar early Pacifics in the Cape. The obvious cleanliness of No. 121 at Sennar in 1944 is worlds away from the sad state of Sudanese locomotives in recent years.**
73. **A long way from home – but not out of place. This former South African Railways 6th class 4-6-0 was photographed at Atbara in 1946.**
74. **Stored before being sold to Rhodesia Railways in 1949, this Garratt was still equipped with wartime headlamp covers at Atbara in 1944. Clouds of sand stirred up by the front unit caused heavy wear on the rear unit resulting in the engines being stored upon arrival of the W.D. 4-6-2s in 1943. At this stage few would have predicted a further career of over 40 years for the Sudanese Garratts.**
75. **Going for nine lives? Lettered for its third owner, CFM, this former Sudanese Garratt is seen on the Beira-Umtali line in Moçambique.**

ETHIOPIA

Imagine a railway which owned more camels than rolling stock, where temperatures frequently reach 70°C (160°F), and where for seven months of the year trains could be engulfed in swirling white dust storms. This was to be found in Ethiopia, 'land of burnt faces', the largest of the countries in the Horn of Africa and at that time loosely-linked with Eritrea whose inhabitants were also Abyssinians. Early-converted to Christianity, and though later surrounded by hostile Moslem states, Ethiopia preserved its independence through its geographical impregnability – an impregnability which was breached by the building of the first permanent railway in the region.

While Ethiopia, to a large extent, retained its independence, surrounding countries were gradually colonized by European powers. To the east and south, along the coast, France, Britain and Italy subjugated Somaliland – fabled 'Land of Incense' – while to the south-west, Britain occupied Kenya and the Sudan.

It was the British who built the first railway in the region, passing through Eritrea, long before Italian rail-construction began. This was the Kumayli Military Railway.

The Ethiopian emperor Theodore had held several Europeans, including the British consul, hostage and in 1867 it was decided to free them by force. British soldiers, commanded by Sir Robert Napier, launched their invasion from Zoualla (Zula), some 55 km south of present-day Massawa. Using second-hand equipment from Bombay they started to build a railway in the general direction of Magdala, 640 km inland and 3 000 m above sea level where the hostages were being held. The Indian equipment sent out happened to be 1 676 mm (5′ 6″) gauge locomotives and rolling stock so that, unwittingly, the

76

76. The daily mixed train to Awash awaits departure at Addis Ababa station in November 1954 – a 1946-built Davenport 2-8-2 leading a 1910 SACM compound 2-8-0, No. 31.

77. From a builder's catalogue: two early Ethiopian engines built by the Swiss Locomotive Works. Only four years separated the construction of the smaller 2-6-0s and the first of the larger 2-8-0s. The latter engines were two-cylinder compounds, having 420 and 630 by 550 mm cylinders, and between 1903 and 1914 a total of 13 were built. The first five were saturated and the last eight superheated.

only example of 'broad-gauge' equipment – apart from some 7′ 0″ harbour construction trackage in the Cape – ever used on the continent was introduced. After six months only 17 km of line had been laid – as far as Kumayli at the entrance to the Soroo Pass – and by then the infantry had marched to Magdala and freed the hostages. The railway was used to withdraw the British forces on the final leg to Zula and, according to the *Book of Firsts,* so became the first to transport British forces in the field. Lacking any further purpose, for the British were happy to leave Ethiopia to the Ethiopians, the railway was uplifted and the six locomotives were returned to India – a mere eight months after the Royal Engineers had begun the rail survey.

As early as 1880, Ethiopia contemplated building its own railway, following an old caravan route from the port of Obook, 15 km from Djibouti, to the capital at Entotto. Nothing came of this until Menelek II, who ascended the Ethiopian throne in 1889, put ideas into action. In 1894, having consolidated his country's borders, he granted a concession to a Swiss engineer and a French entrepreneur to build a line from Djibouti to Entotto and across the Kafta region to the White Nile. However, as Djibouti had been 'French' for several years, he had no right to grant such a concession and legal problems halted the project for two years. When, in 1896, the Ethiopians defeated the Italians at Adowa, the French, possibly hoping to extend their influence in Ethiopia, granted Menelek the use of Djibouti as Ethiopia's official port. Construction of the line began in October 1897 for the 'Compagnie Impériale des Chemins de Fer Ethiopiens' (CIE), and by 1899, grading had reached some 100 km inland – across a waterless, sun-blistered desert of grey volcanic sand. Attacks by primitive, spear-wielding Issa tribesmen added to the railway builders' problems but, in spite of this, a regular service was inaugurated to Km 106 in July 1900.

Competition from caravan operators, who previously had monopolised this ancient trade route, proved fierce and the company was forced to buy them out, adding to its already-heavy financial burden. Nevertheless, the railway pressed on, reaching Dire Dawa 309 km inland and at an altitude of 1 195 m, in December 1902. As this staging-post was reasonably cool, it was decided to establish the railway's workshops here. The railway was careful to operate trains only during daylight as local tribes often ripped up the track at night; and it was not until 1926 that they had been sufficiently pacified so that trains could be run at night.

After reaching Dire Dawa, the railway became embroiled in a Franco-British tug-of-war. The builders were desperately short of funds and, since the French investors seemed unsympathetic, were only too delighted when a group of British financiers formed the Ethiopia Railway Investment Trust and advanced urgently needed cash. This stirred the French into action and a struggle for control of the railway developed with the French finally emerging as victors.

C. S. Small, in his rail-classic, *Far Wheels*, says: 'The British faction, altruistic perhaps through lack of success, felt that since they could not dominate Ethiopia, no one else should. They proposed a tripartite treaty between Great Britain, France and Italy, guaranteeing the independence of Ethiopia. The French, having won the railway battle, were in an expansive mood. The Italians, still smarting from their defeat at Adowa, were in no position to ob-

' This treaty recognised certain spheres of influence of the participants and formally recognised the French Government's interest in a new railroad company which would extend the line to Addis Ababa.

'This treaty now left the French fighting among themselves. The points at issue were the settlement to be accorded to the owners of the original company and the composition of the new one. When a group of Frenchmen get together on such a problem the outcome is inevitable. A private company will be formed using the Government's money and a formula will be devised that insures the promoters a profit with very little risk. Marianne always pays.'

77

It took three years for the parties to agree on an arrangement but eventually the new company was formed with the laborious title,'Chemin de Fer Franco-Ethiopien de Djibouti à Addis Ababa' (CFE).

With its future secure, and reputedly charging the highest railway freight rates in the world, the railway pushed inland. By now what was to be the Ethiopian capital, Addis Ababa, had been established and this became the new objective. The route was quite easy, gaining altitude gradually and staying well away from the mountains to the north – until the Awash River was reached. Here the line had to cross, and then follow, the river's course through a narrow ravine up 1-in-40 grades to a point where it attained the top of the escarpment forming the eastern side of the Rift Valley. From Awash to Addis Ababa was only 100 km as the crow flies, but by rail the journey followed 250 km of winding line. Delayed by the outbreak of World War I, it was completed only in 1917. Since then many projected westward extensions have neared the point of construction. But Ethiopia remains a land of unfulfilled railway dreams.

ERITREA

During the latter part of the 19th century, the Italians, wishing to extend their colonial interests, occupied the coast of Eritrea planning to use it as a springboard for further conquest. As trade and the movement of troops and matériel would have been difficult across the coastal desert and mountains, they constructed a railway from Massàua (Massawa) in 1887, but were so severely rebuffed by the local tribesmen that they were forced to suspend work the following year. A further attempted conquest was again thwarted in 1896, when the Ethiopian emperor, Menelik II, defeated the Italians at Adowa. However, as part of the peace agreement the Italians were allowed to keep Eritrea, and by 1901 construction of one of the most amazing of all mountain railways was resumed.

In 1906, the 'protective' European powers agreed to allow Italy to build a railway linking Eritrea with Italian Somaliland. This line was planned to run west of Addis Ababa and would have been a blatant breach of Ethiopian territorial sovereignty. However, nothing came of this project, nor of a much later and even more ambitious proposal after the Italian conquest of Ethiopia in 1935-36. This plan involved a direct railway line from Massàua to Addis Ababa and on down to Italian Somaliland, with an important branch to the port of Assab. Lesser branches would have extended westwards from Addis Ababa to Ginna and finally one other to Gondor, near Lake Tana.

Nevertheless, Italian railway achievements were formidable. The Massawa-Asmara railway, completed in 1911, is an engineering masterpiece in the best Italian tradition, the 118-km line climbing from sea level to 2 343 m – achieved with a ruling grade of 1 in 28, 30 tunnels, 532 bridges, viaducts and culverts, and with curves of 72 m radius. Less mountainous, but still impressive, was the Asmara-Agordat section, opened in 1922, and the extension to Biscia, opened in 1930 but now closed.

Farther south, along the coast at Mersa Fatma, a 65 km 'Decauville' (600 mm) line was constructed inland to Kolulli. Five 0-4-0T locomotives built by Decauville operated on this line, which also had two small O & K and Henschel tank engines and three American Porter 0-4-0Ts, built from 1900 to 1919. When this line was closed in 1929, most of the locomotives seem to have been scrapped, although the Porters escaped the torch by being transferred to Italian Somaliland in 1924 where, re-gauged to 950 mm, they continued their careers.

The last railway built in Eritrea was very short, in both length and lifespan. It was located in Assab, the southernmost port in the country, which had been bought from the local Arabs by the Italians in 1869. Although the nomads' ideas of land ownership must have differed markedly from those of the Italians, nobody disputed their right to this godforsaken piece of real estate; they dreamt of the day when Assab would become the major port for southern Ethiopia – in direct competition with French Djibouti.

In the late 1920s the Italians built a tarred road through to Addis Ababa causing the French-owned CFE (Djibouti-Addis Ababa) railway considerable grief. Finally, in 1939, the port was improved for strategic reasons and a locomotive was brought in to help with the construction. This was an old 1886 R.W. Hawthorn & Leslie 0-4-0T, which had operated on a mineral railway in Sicily as an 850 mm gauge locomotive but was regauged to 950 mm for Eritrean use. In the early 1950s C.S. Small found not a trace of the railway although road transport operators were still doing a flourishing business with Addis Ababa, no doubt to the displeasure of the CFE railway.

78

79

SOMALIA

The last of the Italian railways were located in Italian Somaliland, which as present-day Somalia has no railways of any kind. In the 1920s, Italy promoted immigration to its colonies; in Somaliland several farming communities were established, including Villaggio Duca degli Abruzzi, some 50 km inland from Mogadiscio. This small village with a big name had grandiose ideas of development to match its ducal title. With the support of the colonial Governor, apparently a close relative of the Italian king and the man after whom the village had been named, a 950-mm gauge railway was built. It started at Mogadiscio on the coast and worked its way over low hills into the Shebeli river valley eventually arriving at the town 113 rail kilometres later. The line was opened in 1927 and began operation with a number of old steam engines, all of which seem to have come from other railways – but in 1928, of seven locomotives said to have been on the roster, only one was operational. To solve this problem some railcars and two diesel locomotives were obtained, but these only increased the already severe difficulties; this railway must have been the most unreliable ever. By 1934, however, additional diesel locomotives had been bought and the last mainline steam locomotive, an 0-6-0T of class R 301 was transferred to Eritrea, while the three old Porter 0-4-0Ts which had originally operated on the Mersa Fatma line in Eritrea were retained for shunting at Mogadiscio.

At least two 600-mm gauge Decauville lines existed in Somaliland. The first, which was built as part of an agricultural development project at Genàle, a short distance south of Mogadiscio, came into being before 1920, making it the colony's first railway, but it was abandoned in the mid-1920s and little is known of its locomotives except that steam was used first and diesels later. Some of the equipment may have found its way to the next Decauville project, which was also initially intended for agricultural purposes. This was the Villaggio Duca degli Abruzzi to Bulo Burti railway, a 130-km line which was reportedly taken over by the military authorities for service in the Abysinnian campaign, converted to 950 mm gauge and extended a farther 150 km to the border at Ferfer. The British forces captured Mogadiscio on February 25, 1941, and soon controlled the entire railway. Within a year they decided that the railway would be more useful elsewhere in the war campaign and everything was disassembled and sent to widely dispersed destinations.

So history absorbed the Italian Somaliland railway which, had the grand plan of its promoters been realized, would have formed part of the 2 200 km line linking it with Eritrea.

The Italians fared no better against the Allies in Ethiopia and Eritrea. In spite of some early successes, by March 1941 the Italian army was retreating and within a month the entire Eritrean railway was in Allied hands. In Ethiopia, the Allied forces advanced to Dire Dawa and on to the Awash River – to find that the Italians had destroyed the major bridge. Within six weeks, the South African Brigade built a completely new rail bridge and by early May Addis Ababa was occupied. The Italian political presence had ended, though in Eritrea today her cultural influence is still noticeable, the cuisine's emphasis being on pasta, olive oil and garlic.

Since the colonial powers departed, the entire region has seen much political upheaval. Ethiopia was recognised as an independent state in January 1942, but Eritrea was a United Nations mandated territory until 1952 when it was federated with Ethiopia. This move angered many Eritreans and by 1971 the Eritrean Liberation Front had begun an armed struggle which severely disrupted railway operations. Along the border with Somalia, which became independent in 1960, old grievances erupted and by 1977 a full-scale war was being waged. With the French withdrawal from Djibouti in 1978 (formerly French Somaliland) the situation is even more volatile.

Locomotives

The sparsely-populated countries of this region have only operated some 1 400 km of railways, with less than 1 100 km of line in operation today. Traffic has never been particularly heavy and, in fact, road transport has held the edge for many years. The Massawa-Asmara line, with its exceptional

80

78. **In November 1954 C.S. Small photographed the locomotive shed at Addis Ababa where he found No. 26, a compound 2-8-0 (low-pressure cylinder clearly visible) having a tube-clean. Seen behind, is one of the railway's, light 4-6-2s.**
79. **Italian and American locomotives at Addis Ababa – the 'Yank' product – unassuming and successful; the Italian impressive in appearance – but unsuccessful in operation.**
80. **No. 233, a 1937 Haine St. Pierre 4-6-2 – note the coal briquettes in the tender. Ethiopian engines burned coal even though a ready source of liquid fuel was available just across the Red Sea.**

grades, could handle only 2 500 to 3 000 gross tonnes a day, so the Italians built a well-designed, parallel highway and an aerial-bucketway which is a marvel of engineering expertise although not in use today.

Over the years only about 200 steam locomotives have graced the rosters of the seven different railways which operated in the region. The Djibouti-Addis Ababa railway (CFE), with the longest mainline, had the largest roster – 96 steam locomotives throughout its existence. In Eritrea, the Massawa-Agordat line was a close second with 86, but as its mainline was only 40% of the length of the CFE's, it had a greater concentration of motive power.

Understandably, the choice of locomotive builders was dictated by political considerations. Italian builders supplied 84 engines, 76 of which operated on the Asmara line; French builders supplied 64 locomotives, 59 of which operated on the CFE; Swiss, Belgian, German, British and American builders also contributed, but in a smaller way. Overall, 'continental' builders supplied more than 90% of the region's needs and, as might be expected, two very 'continental' locomotive types appeared: the compound Mallet tank engine and the two-cylinder compound 2-8-0. The Mallets arrived first, becoming the mainstay of the Asmara line, while the two-cylinder compounds were the forerunners of a long line of superheated, simple-expansion 2-8-0s operated on the Addis Ababa line.

When first built, this railway had a gauge of 750 mm and seven small 0-4-0Ts were built by Henschel in 1887 for use on construction work. Immediately before construction was resumed in 1901, when the line's gauge was made to conform with that of Italy's own secondary railways, these small engines were converted to 950 mm gauge.

By 1904 the railway had crossed the dry coastal plain and the first mountain foothills to Ghinda, 69 km from the coast, and was poised for the assault on the mountain escarpment guarding the approaches to Asmara. For powerful haulage capabilities in such terrain, articulated locomotives were needed and in 1907 the first Mallets arrived: small 35-tonne 0-4-4-0Ts typical of the period. They proved an immediate success and between 1911 and 1915 Ansaldo of Genoa built 25 duplicates. Three more engines were constructed from spare parts in 1931 and 1938, bringing the total in class 440 to 31 engines – two of which were still operating in the early 1970s.

The next Mallet type, class 441, was considerably larger, weighing

81 82 83
84 85

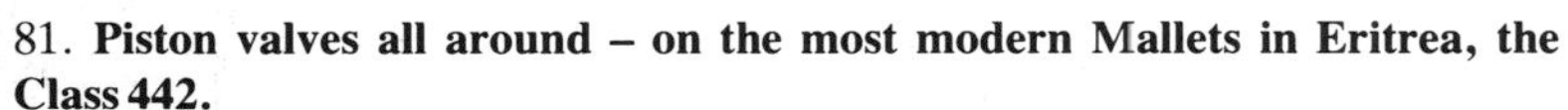

81. **Piston valves all around – on the most modern Mallets in Eritrea, the Class 442.**
82. **An older, slide valve 440 Class No. 021 storms through the streets of Massawa, past the Sheikh Al-amin school, as it heads out of town in February 1970.**
83. **No. 202-007 built by Breda in 1937 with a short train for the Melotti Brewery, in the suburbs of Asmara.**
84. **Amid piles of parts including a steam dome with a large Ansaldo plate, this Mallet awaits repair at Asmara.**
85. **In March 1967 Jeremy Wiseman, a passenger on the Massawa-Asmara train, captured this record of its tortuous uphill journey.**

202
007

46 tonnes and differing from the class 440 by having single-expansion steam distribution. Sixteen of these were built between 1933 and 1936 and were the most powerful engines to operate on the system – having a tractive effort of 13 606 kg (29 970 lbs) compared to 6 946 kg (15 435 lbs) of the class 440 Mallets. The first eight 441s, delivered in 1933, included three engines with Walschaert valve gear and five with Caprotti valve gear. These engines were all super-heated, but the second batch, delivered in 1936 were saturated, although having Walschaert gear. Later, one of the final batch was converted to compound working, while in 1945 four others were transferred to Libya by the British administration. Towards the end of the 1950s, only the single compound engine was still in service in Eritrea – an indication that the Italian builders possibly miscalculated the boiler's ability to supply enough high-pressure steam to the four hungry cylinders of the 441 class engines.

The final class of Mallets, the 442s, were all compound and superheated. Eight were built in 1938 and though they weighed slightly more than the 441s, their tractive effort, because of compound working, was naturally less (11 400 kg/25 140 lb). They were the most successful of the 54 Mallets operated and it is worthy of note that as late as 1963 an additional engine was built up from spare parts. The nine members of this class were still in service in the early 1970s, being the principal mainline engines on the mountain section. Sadly, recent reports indicate that several have been shunted onto viaducts and dynamited – victims of Eritrea's civil war.

In view of this railway's reliance on articulated power, the absence of Garratts is surprising, but the Italians, like their continential neighbours, had a strong pro-Mallet bias. (When the Italians finally did attempt to build a Garratt, the results were not a success.)

Between 1942 and 1952 the railway was administered by the British and it has been reported that a British railway official once remarked: 'A properly designed Garratt should pull at least twice the load of the toy engines used by the Italians.' However nothing was done as the British probably regarded the railway as too unimportant to merit improvements. Anyway, the introduction of more powerful locomotives would have created other problems, for the numerous passing loops on the mountain were designed to carry trains limited to 12 vehicles weighing 100 gross tonnes. Such braking as there was also presented problems. Through train-line brakes had never been used and, as a result, two brakemen were required for every three wagons . . . to set the brakes by hand as the train rolled downhill.

86. **A unique locomotive, this Klien-Lindner 0-8-0T built by Ansaldo in 1922, originally operated in main-line service, but in later years she was restricted to shunting at Massawa.**
87. **In the mountains. A typically well-constructed Italian viaduct supports the mass of the westbound mixed, a 442 Class compound Mallet providing power. Close scrutiny of this impressive photograph reveals no less than three tunnel portals in the mountainside across the valley.**

86 87

88

89

88. **Above the clouds and near the summit of the climb, different levels of the spectacular mountain section of the Massawa-Asmara railway are seen, as is the steepness of the gradient on this line.**
89. **Heading for the mountains. Two older 440 Class Mallets doublehead across the semi-desert coastal section near Mai Atal in November 1954.**

The most remarkable locomotives to operate in Eritrea were five Klien-Lindner 0-8-0Ts, built in 1922. They coincided with the opening of the Asmara-Agordat section and probably were purchased as an alternative to the Mallet. In service they were found to be rather hard on the frail track and eventually were relegated to lesser duties. Four remained on the roster in 1952 but when C.S. Small visited the line a few years later, all he could find were their skeletons near the Asmara shed.

For shunting, 11 ultra-small 20-tonne 0-4-0Ts were bought between 1927 and 1937. At last report seven still operated, but this was before strife enveloped the area. Other non-articulated engines acquired came from various sources and included 2-6-0Ts and 0-6-0Ts, all obtained second-hand from Sicily and Somaliland.

This line was early to use diesel traction, introducing railcars in the 1930s. These 'Littorinas' cut the travel time from Massawa to Asmara from ten hours for steam to a mere four hours. Unfortunately, in later years, they fought a losing battle with bus operators on the adjacent highway. Other than using some small, ineffective diesels obtained from the Mogadiscio line, the system operated entirely on steam for road service until 1957 when three 650 hp diesels arrived from Krupp. Though these Bo-Bo diesels were intended for the mountain section, they were later transferred to the less arduous Asmara-Agordat stretch where they have remained.

In the mid-1970s, before civil strife reached disruptive proportions, the railway had 18 steam engines, three diesel shunters, three diesel mainline locomotives and five railcars. The line's name had been changed to the Northern Ethiopia Railway Share Company and the annual *Directory of Railway Officials* – that useful yearbook which details locomotives and equipment of the world's railways – gave no indication of anything amiss. But the reality is very different and it is likely that one casualty of the conflict will be the railway, particularly its steam locomotives.

Where Italian influence was the hallmark of the Eritrean and Somali networks, it was the French who influenced Ethiopia's railway. The first locomotives were French – four small 0-4-0Ts built by Gabert-Frères of Lyons in 1898 for the Compagnie Impériale. These were numbered 1 to 4 and were given names, as were all following Ethiopian locomotives. By 1907 the original engines had been renumbered 51 to 54 and in 1910 four more identical locomotives, numbered 55 to 58, were built by A. Pinguely.

For road service a larger locomotive was needed but, with the railhead still far from the mountains, a Mogul type was considered big enough. In 1899, six 2-6-0s came from the Swiss Locomotive Works (SLM) and a repeat order was placed in 1901, but these engines were lost at sea. Although three replacements were built in 1912 these light, 29-tonne saturated engines struggled, even on the fairly easy grades between the coastal plain and Dire Dawa, and as early as 1903 an order had been placed for 8-coupled machines.

Again, SLM gained the contract and built four 34,5-tonne engines, numbered 21 to 24. These were Ethiopia's first two-cylinder cross-compound locomotives, then very much the vogue on the mountain railways of Eastern Europe. These engines were the last bought by the Compagnie Impériale. The next order, in 1910, for additional engines of the same type but superheated this time was for the new company, CFE, and the stronger interests of the French company saw to it that SACM was awarded the contract.

The next locomotives were simple expansion and super-heated. These comprised nine 2-8-0s, built by SACM in 1913 and were destined to become the largest class on the system. Similar in size to the compounds, additional batches of these simple engines continued on the order books until 1938 and 30 2-8-0s were eventually in service.

A heavier Consolidation type with wide firebox was ordered in 1927 and, over a decade, 13 of these 47-tonne machines were bought. Finally, in 1938 the railway acquired six 2-8-0s, similar in design to their own first simple 2-8-0s. These engines had been built by SACM in 1908 and 1910 as tank locomotives for the Appenzeller-Bahn in Switzerland and had been displaced by electrification. This gave the railway 62 sure-footed Consolidations, more than 75% of its entire locomotive fleet.

With this substantial freight power, the railway looked for more elegant engines for its passenger trains. In 1936 it found a lovely 4-6-2, built for the Madrid-Aragon Railway in 1915 but never delivered and stored in the works of Haine St Pierre. With this in service it was decided to order three more like it and these were built in 1938, being numbered 232 to 234. These engines had the largest coupled-wheels (1 250 mm) of any locomotive on the roster and, weighing 48 tonnes, were even heavier than the 2-8-0s.

By the late 1930s all the region's railways were under Italian control, giving Italy's locomotive builders a chance to match their designs against those already in operation. But they failed. In 1938 Ansaldo, the builder who had constructed many of the Mallets for Eritrea, built six 2-8-2+2-8-2 Garratts. Only three reached Ethiopia, two being diverted to the Tripolitanian Railway – where they were never used because of war damage – and one lost at sea. The three which reached Ethiopia were put to work, but apparently their design was so deficient that they spent most of their time in the repair shops. When C.S. Small visited the railway in the mid-1950s only one was intact – rusting quietly at the Addis Ababa engine shed. A sister engine had donated its boiler to a nearby factory while the third had vanished.

CFE's last steam locomotives were delivered in 1946 and two quite contrasting designs were acquired. The first were six sturdy U.S. War Department design 2-8-2s, built by Davenport, while the second type was a single 1917 Henschel compound 0-6-6-0T Mallet. This engine, one of the standard German 'departmental' design widely used in World War I, had enjoyed a varied career. In 1929 she travelled to Switzerland and was used on the Yverdon-St Croix Railway until electrification. When she arrived in Ethiopia she was reportedly in good condition but was seldom used before being scrapped. Probably she was just too slow, but her presence meant that for a short time the railway possessed both the Mallet and Garratt types and, it can be concluded, was impressed by neither.

Any railway running steam locomotives through exceptionally dry country, as does the CFE between the coast and Dire Dawa, experiences water problems. So it is not surprising that diesels made inroads quite early – the first arriving in 1950, when the railway still had 90 steam engines on its roster. By 1955 12 diesels were handling all services across the desert to Dire Dawa, while a mixed steam-diesel operation ran to Awash. Beyond, to Addis Ababa, steam remained supreme, but only for a short time – between 1956 and 1962 some 19 additional diesels arrived and, as early as 1958, the railway could report that steam handled a mere 1,2% of the traffic. In spite of this, the last steam was only officially retired in 1965, although it had been quite inactive towards the end.

That few rail-photographers visited this operation is unfortunate, for spectacular scenery is encountered between Awash and Addis Ababa. The area is very much off the beaten track, far from the usual tourist routes, and though steam enthusiasts have roamed the globe in recent years, only C.S. Small visited and recorded steam in the area. He travelled these tracks long before other enthusiasts and we feel it appropriate to end this chapter with a quotation from his *Far Wheels* which may well evoke similar memories for other rail travellers:

'The Awash hotel must be seen to be believed. There are two rooms with brass bedsteads equipped with mosquito nets for the occasional first-class passenger or railroad official. A series of cubicles serve the less fortunate who arrive on the third-class cars. The number of flies is astounding. At breakfast or dinner when the flies are most active, eating is a race between the diner and the insect. Dead flies in the marmalade or floating in the soup, and live flies darting into one's mouth constitutes the acid test for the collector of antique railroads.'

4 EAST AFRICAN RAILWAYS

Second only to South Africa in route kilometrage, most of the EAR lies south of the equator yet, in various guises, it 'Crosses the Line' no less than five times. Only two other railways in the world make the trans-equatorial crossing – in Zaire and Ecuador – each only once. A third line, in Sumatra, which also held this distinction, no longer operates. In one of the five crossings the EAR main line reaches an altitude of 2 783 m (9 136 ft) at Timboroa – the highest of any stretch of rail in Africa and, in colonial days, the highest in the British Empire.

Unlike most other British-built railways in Africa south of the equator, the Kenya-Ugandan lines were built to metre gauge rather than the usual 3′ 6″ – a fortuitous development, for the German-built lines in Tanganyika were also metre gauge, though at the time there can have been no thought of common ownership or operation. However, this fortunate co-incidence of gauge greatly simplified the eventual incorporation of the two systems.

With so fascinating and challenging a patchwork of lines it is not surprising that the range of locomotives employed was itself spectacular. Just as South Africa used the largest and most powerful locomotives ever to run on 3′6″ gauge, so the corresponding behemoths of metre gauge were to be found in East Africa.

GERMAN EAST AFRICA
(now Tanzania)

Surprisingly, in view of their tardiness as railway-builders in other colonies, in East Africa the Germans were ahead of the British . . . but then they did not have to contend with a British Parliament which was to discuss for years the pros and cons of a railway in 'Darkest Africa'. Three German lines were built – one was a 750 mm-gauge railway that was closed in 1924, a scant 12 years after its completion – and the others, metre-gauge railways that were not linked as an integrated rail system until 1955.

Of the three, the Usambarabahn, named for the Usambara Mountains which it paralleled, was the first; as Tanganyika's northernmost railway and because of its proximity to the Kenya border it was probably as important strategically as it was commercially. The first track was laid from the port of Tanga in 1893 and progressed over fairly easy country, but failed financially after only 40 km had been built. Work was resumed in 1899 and its first terminus at Moshi, nestling against the foothills of Mount Kilimanjaro, was reached in 1912. The Germans planned to push the line as far as Lake Victoria but, probably as a result of the rugged terrain, the line was taken no further than Arusha, which is still the terminus.

After World War I, when Britain took control of what then became Tanganyika, a link with the Kenya-Uganda railway was made at Voi, but it was not until 1955 that the old Usambarabahn was connected with the Tanganyika Central line, underlining, perhaps, its comparative unimportance.

The locomotives of the Usambarabahn were, by and large, scarcely more impressive. The first five were 0-4-2Ts built by Vulcan, of Stettin, and these were used for track-laying and the initial train services. Only spark-arresting chimneys and enlarged bunkers for wood fuel distinguished them from the numerous similar engines used on the secondary railways, or 'Kleinbahnen', of the Fatherland. Even the square-cornered numberplates mounted on the smokebox sides were copied directly from the Prussian State Railways. For seven years these five diminutive 19-tonners handled the entire traffic of the infant line.

In 1900 Jung built five 26-tonne 0-4-4-0T Mallets which, with roughly double the Vulcans' haulage capacity, augmented these earlier models. Mallet tanks of this type were becoming quite common on the little railways of Europe and were thus an obvious choice for the German colony. These Mallets were East Africa's first articulated locomotives and it is unlikely that anyone connected with them would have foreseen the development, slightly more than half a century later, of an East African articulated locomotive ten times the mass of this early Jung design. Eight years were to elapse before the next batch of motive power was acquired – in the form of four German colonial standard 2-8-0Ts, built by Orenstein & Koppel. These followed a basic design – but with varied detail and dimensions, and were used extensively in most of Germany's African territories. Only in one other sphere was such extensive use made of the 2-8-0T, and that a far remove from colonial Africa – in the South Wales coal-hauling railways.

The same German suppliers built four 2-8-0s between 1910-12, virtually tender versions of the previous tank engines, though with larger boilers and increased fuel and water supplies. These were the last locomotives built for the Usambarabahn which, after World War I, became part of the British Tanganyika Railway.

However, the last two German engines to run in East Africa, a pair of 2-8-0Ts from the Central line, finished their lives at Tanga in 1950 – thus the launching point of the Usambarabahn was both birthplace and graveyard of German locomotives in Tanganyika over a life-span of 57 years.

The Ost Afrikanische Mittellandbahn, or Central line, was operated out of Dar es Salaam by the Ost Afrikanische Eisenbahn Gesellschaft – East African Railway Company – a somewhat grandiloquent title, as it did not control all the systems in the German colony, let alone East Africa as a whole. Although mooted in 1891, building of the OAEG line was not started until 1905, some 12 years after the Usambarabahn. Nevertheless, this was by far the most important system and by 1914 the line had reached Kigoma, on the shore of Lake Tanganyika, having traversed the entire width of the colony.

A branch of this line pushed out from Tabora to Mwanza on Lake Victoria; this proved to be a shorter and less difficult route than the proposed extension of the Usambarabahn. Other branch lines were built to Mpanda and Kidatu. With lighter traffic and easier gradients than on the British line through Kenya, fairly small units of motive power were always adequate.

The first locomotives introduced were fairly similar to those used initially on the Usambarabahn. There were four 0-4-0Ts built by Henschel in 1905 and another four in 1909, all used for construction work. A second-hand, 0-4-0T, built by Markmaschinenbau in 1893, arrived between the two batches from Henschel and was probably used by the railway building contractor and then taken over by the railway.

To start the line service, small Mallet tanks were supplied, the first, as on the Usambarabahn, being Henschel-built 0-4-4-0Ts – four in 1905 and another in 1907. These were enlarged into 2-4-4-0Ts, to improve tracking and increase fuel and water supplies, Henschel building four in 1908. After this, the Mallet was abandoned in favour of the better-designed colonial 2-8-0T. The Mallet was probably more suitable for construction tracks before the formation had consolidated, and doubtless was still used at the railhead sections. A total of 24 standard 2-8-0Ts were supplied during 1909-10, by Henschel, Borsig, Maffei, and Orenstein & Koppel, all to basically the same design. Henschel also built two 0-8-2Ts for comparison but, not unexpectedly, these did not track as well, owing to the absence of leading trucks.

As the line progressed and tender engines became necessary, OAEG ordered 20 Hanomag 2-8-0s, which were supplied between 1911-13. These were the last new engines ordered under German auspices, the British replacing them fairly speedily by a motley collection of ex-Indian or Indian-inspired locomotives. This was a blow to the enthusiast, for R. Ramaer recently unearthed diagrams of several German builders' proposed designs for OAEG, some of which would have been very interesting if built. The 4-8-0, 2-8-2, and 4-8-2 designs were of much the same size as those subsequently built, although more modern in appearance, while there was also an amazing 4-10-0 proposal, which would have baffled British colonial designers.

90. **Africa's Great Rift Valley, one of the continent's physical wonders and a major obstacle to the builders of the original Uganda Railway, dwarfs the Gilgil to Nairobi local goods, hauled by a 2-8-2 No. 2930 'Tiriki'.**

THE UGANDA RAILWAY

To the late 19th century Englishman, Uganda epitomised the romance of 'Darkest Africa'. Within its ill-defined boundaries the mysterious Nile rose to begin its long journey northwards to the Mediterranean; huge tusks of ivory were shipped back to Britain by hunters and traders who threatened to wipe out the herds of elephant which shared the land with 'savage' tribes. It also provided a rich harvest for the Arab slave traders, operating from Mombasa, who marched their captives loaded with ivory across the intervening 800 km of wild bush country which was to become known as Kenya. Mortality rates in the slave coffles were high, but the slavers' rewards for those captives who survived the gruelling trek were even higher.

To the philanthropic Victorian, such decimation of the indigenous tribes had become a clarion call to action. The slave trade must be halted and, in the absence of navigable waters for the traditional gun-boat, a railway would carry the flag. The proposal did not have an easy passage through Parliament, as attested by reams of flowery verbosity in *Hansards* of the time. The main opposition came from the core of hard-headed financiers to whom the construction of a railway through more than 800 km of desolate wilderness represented the ultimate folly. The debate dragged on for years before the pro-railway lobby triumphed.

As railway building through hostile terrain, the Uganda line must rank in the world's 'big three'. The coastal lowlands were rife with malaria, along with other diseases, and water shortages plagued them across Taru desert. Lack of labour was a major obstacle and, as the local tribes had been so weakened by the slavers, coolies were brought in from India. Yet problems were by no means at an end. Around the Tsavo River work was disrupted by man-eating lions, among them a pair which terrorised the construction camps for nine months, claiming 29 victims before they were shot by one of the engineers. Later, crossing the Mau escarpment, and then dropping into the Great Rift Valley the topography presented almost unsurmountable problems and hostile tribes had to be placated.

In spite of these difficulties, the 584 miles of railway from Mombasa to Port Florence (now Kisumu), on Lake Victoria Nyanza, were completed in the incredible time of six years – from 1896 to 1901. Today, with modern earth-moving equipment, and without the hazards encountered 80 to 90 years ago, few companies could execute a similar contract in the same period.

By the time it was decided to build the Uganda Railway, Cecil Rhodes' dream of a Cape-to-Cairo railway was being widely canvassed and it was clear that the new line would play significant part in the grand plan. With the 3′6″ gauge already established in South Africa and proposed for the Sudan, the decision to use metre-gauge track in Uganda seems an example of bungling at its worst. In fact, it was a matter of expediency, for engineers, material and labour for the line were all brought from India where metre gauge was well entrenched . . . for secondary railways. A misunderstanding had led the planners to believe that Sudan would build to metre gauge. But the Uganda Railway was to be a main line and it is surprising that those with the foresight to insist it should be built did not see that it should harmonize with the other systems likely to connect with it.

Lumbered with metre gauge from India, it was inevitable that the Uganda Railway would be provided with whatever surplus metre-gauge locomotives were available. India was not to blame for these shortcomings of its metre-gauge motive power. On the Deccan, metre gauge was suitable for secondary railways. To assume this equipment would be adequate for what was virtually an international main line traversing exceptionally severe terrain, was expecting too much. The A class 2-4-0Ts were so limited that they never ventured from the harbour area. The E class 0-4-2 tender engines were equally feeble but had a longer range, enough to propel a truck or two of needed supplies, sleepers, curry powder and beer, to a not-too-distant railhead. When it came to the F class 0-6-0s, and the N class 2-6-0s, India gave of its best, and their inadequacy was no fault of the suppliers.

Probably the finest of the early engines were the B class, Baldwin Moguls, built in 1899-1900. Though nominally of similar dimensions and no more powerful than the F class 0-6-0s, their flexible wheelbase with compensated springing and leading pony truck made them more suitable for rough pioneer track. Had South African 3′ 6″ gauge been adopted, the sturdy, eight-coupled engines of either Natal's Dübs A, or the Cape's 7th classes, would have been available, either new or second-hand, and these would have been far more useful than the Indian types.

After the B class, no further engines were acquired until 1912-13, when steadily increasing traffic demanded more locomotives of much greater power. For freight work, North British built a pair of 0-6-6-0 Mallet tender engines, each of approximately double the tractive effort of the F class. They used saturated steam, and all cylinders had slide valves, the general design being more or less a North British standard as supplied to India, Burma and Spain. A further six in 1913 had piston valves on the high pressure cylinders, as did a final ten built in 1914. In service they pulled well, but rode roughly and were heavy on maintenance and so were rapidly, and thankfully, replaced as soon as Garratts became available. There were also two classes of tank engine, both by Nasmyth Wilson. Three were 2-6-2Ts of more or less standard Indian design, and eight were basically the same engine enlarged to a 2-6-4T, with more fuel and water. For a time these were used with auxiliary tanks on main-line passenger turns, being slightly more powerful and probably much better riders than the tender engine alternatives. After being replaced on main lines by 4-8-0s, they were relegated to the shunting and branch working, for which they had been intended.

The final few years of the Uganda Railway were characterized by the acquisition of numerous 4-8-0 tender engines, all basically to the Indian BESA designs, but developed to suit local conditions. From 1914-19, 41 of the slide valve, unsuperheated classes EB and EB1 by Nasmyth Wilson and North British were bought, and many of these survived into the 1950s on lightly laid branches. By contrast, the two experimental superheated, piston-valved engines, class EB2, by Nasmyth Wilson in 1921, had quite a short life, although they justified their existence by proving the advantages of superheating. From then onwards, all main-line engines were superheated. The EB2s led directly to an enlarged version with improved cylinder design, class EB3, of which 62 were built – 56 by Vulcan Foundry in 1923-26, and a further six by Nasmyth Wilson in 1930. These handy and surefooted engines have always been popular and were the most numerous class used in East Africa. Even today, it is reported that they are kept in a strategic reserve, while more modern and powerful steam classes are being withdrawn for scrap.

THE KENYA – UGANDA RAILWAY

By 1926, Kenya's growing importance could no longer be served by the accidental passage of the Uganda Railway, and the railway's name was changed accordingly. Traffic was increasing rapidly and more locomotives, of greater power, were needed. The first additions were an improved shunting class, similar in size to the 2-6-4T, but with wheel arrangement altered to 2-6-2T with more adhesion mass. Between 1926-29, 27 were built, the earliest lettered UR, and some remained in light shunting service to the end of steam.

For main line work, Robert Stephenson built six engines comprising the magnificent EA class of 2-8-2, very substantially more powerful than the 4-8-0s, with double the grate area, 63% more tractive effort, and 75% more adhesion mass, an almost unprecedented increase in capacity. Their overall size and power were closely comparable to Gresley's P1 class heavy freight engine in Britain, and when built were the largest and most powerful non-articulated metre-gauge locomotives in the world. In some ways they were perhaps a little too big, for they suffered from rough riding and hot axle-boxes, later cured by fitting roller bearings. Splendidly handsome, they were given local names, and because of their high axle load, were at first confined to the Nairobi-Mombasa main-line, though eventually they were allowed to Nakuru after this upcountry section had been strengthened. The EAs were the last non-articulated main-line engines built for the KUR, for the Garratt was proving an unqualified success.

The first four Garratts, class EC, were a logical development of the highly successful EB3 class 4-8-0, and in most mechanical details were standard with tender engines and pony trucks being added to the inner ends of each frame, making them of the 4-8-2+2-8-4 type, the first of this popular wheel arrange-

91. **Its antagonists derisively nicknamed the Uganda Railway 'The Lunatic Line'. In this old photograph an F Class 0-6-0 is winched by cable over temporary track, laid while the main line was being engineered down the Great Rift Valley. With the wisdom of hindsight, it is clear that the 'lunatics' were those who opposed building this now important railway.**

92. **Typical of the earlier motive power on the German OAEG in Tanganyika, this Henschel 2-4-4-0T Mallet would have been equally at home on a secondary railway in the fatherland.**

93. **Before the introduction of Garratts, Kenya's first venture into articulated motive power was a batch of 0-6-6-0 compound Mallets by North British. A builder's employee poses beside the finished product.**

94 (Overleaf). **Many Sikhs became drivers on the EAR, and were particularly proud of their charges. In his cab, decorated with personal clock, and brass elephants over the firebox door, Mr Kirpal Singh Sandhu sits at the control of No. 5918 'Mount Gelai'.**

93

ment to be used anywhere. A large boiler with Belpaire firebox was slung in-between, providing more than double the capacity of 4-8-0s, and ensuring ample steam supplies. Two further batches were built, making 36 of the same basic design produced from 1926 to 1931. The second batch was classed EC1, built in 1927, and two more of this class, but modified front ends, followed in 1930, one of which had an ACFI feed water heater. All of these were built by Beyer Peacock, but the final ten EC2s were supplied by North British in 1931.

All these early Garratts lasted well, though six sold to the Yunnan railway in Indo-China in 1939 have not been heard of since. The others soldiered on for about 30 years, being withdrawn from service between the mid-1950s and early 1960s, their last duties being on branch lines and on heavy shunting and transfer service. They demonstrated the superiority of the type, and whereas the EA class 2-8-2s needed heavy rail to run on, the Garratts were capable of handling the same load on light rail.

The final pre-war Garratts ordered for the KUR were the remarkable class EC3, whose brilliant design merited several accolades – they were the world's first 4-8-4+4-8-4 design, the heaviest and most powerful locomotives to run on 50-lb track, and at the same time impressively handsome. Beyer Peacock built 12 in 1939-40, and they proved invaluable in moving the heavily increased wartime traffic. Apart from being large and powerful, they also had the largest wheels used on the railway, so adding speed to their mountain climbing abilities. Although all are out of service today, one of these remarkable machines has been preserved in the EAR museum at Nairobi.

East Africa occupied a strategic position in World War II, mainly as a staging point for men and supplies en route to the battles of North Africa. As a result, it was one of the few regions to which Britain exported locomotives during hostilities, other lines making do with their existing engines. Garratts were produced for the War Department, based for expediency on existing designs, and of these, the KUR was allocated seven of the heavy 4-8-2+2-8-4s and two of the light design of the same wheel arrangement, classing them EC4 and EC5 respectively. The EC4s were used only on 80-lb track and for ten years were the most powerful locomotives on the system.

After the war two further designs of Garratts were ordered. First were 18 more of the highly successful class EC3, modified with slightly larger cylinders, stovepipe chimneys and piston tail rods.

By now, the restricting Indian height-loading gauge had been raised, and these locomotives were correspondingly taller, presenting a less impressive sight than the first batch. The other new locomotives were a batch of 4-8-2+2-8-4s built for use in Burma, where the unsettled political climate prevented their delivery, but which were welcome on the EAR, particularly as they were largely standard with the class EC5, from which they had been developed. The ex-Burma engines were class EC6, and the first in East Africa with the streamlined front tank originated by Beyer Peacock.

95

96

99

POWER IN REPOSE

These portraits of East African locomotives show some of the wide range of engines which were used on the various lines of this system.

95. **Light 2-6-2T shunter No. 1126 in ex-works condition.** 96. **The final light shunters, Class 12 2-6-2Ts for Tanganyika are exemplified by No. 1201.** 97. **Tanganyika Central Line 2-8-4 No. 3019.** 98. **A 4-8-4+4-8-4 Garratt, No. 5804, at Kisumu shed in 1971, awaiting a turn of duty to Nakuru.** 99. **This 4-8-0, No. 2453, outside Mombasa shed in 1975, is dwarfed by its 59 Class stablemates in the background.** 100. **A Tanganyikan 4-8-2, No. 2106, gleams in the sunshine.** 101. **Tanganyika's British-built 25 Class 2-8-2s were a realization of an earlier German proposal for this type.** 102. **A Class 26 2-8-2 at Tabora in 1976. Developed from the earlier 25 Class.** 103. **The last steam engines to arrive in Tanzania were second-hand Indian YG Class delivered in 1978.**

101

97

98

100

102

103

KENYA'S BIGGEST GAME

104. **As East Africa's largest locomotives, and 'double-mountain' types themselves, it was appropriate that the 59 Class were all named after local mountains, with altitudes shown as an interesting detail. Close-up of plate on No. 5902.**
105. **The lineside mosque at Mackinnon Road commemorates the death of a holy man eaten by a lion during construction of the Uganda Railway. Frequently gigantic 59 Class Garratts, thundering past on freight trains, would disturb its peace.**
106. **Big game at Voi – a 59 Class accelerates towards Nairobi past a typical acacia tree, with the Taita Hills as a backdrop.**
107. **Low light and stormy weather outline the bulk of a 59 Class standing in a typical Kenyan wayside station.**
108. **Climbing through rugged country nearing Nairobi, No. 5910 'Mount Hanang' approaches the lonely station of Kiu.**

E A R

109

110

LESSER GAME

109. **Although East Africa is often thought of as an area where the Garratt reigned supreme, in fact non-articulated locomotives outnumbered articulated ones, both in numbers and classes. Post-war EAR operated 11 Garratt classes and 16 non-articulated classes (including four tank-type). Of slightly more than 600 metre-gauge steam locomotives which have operated in the region, 496 came to or were purchased by EAR, comprising 149 Garratt, 59 tank, and 288 tender types. Typical of the latter is No. 3120, bowling through the dry scrub area of eastern Kenya near Ulu.**

110. **A Tribal Class 2-8-2, No. 2919, accelerates out of Mnyusi with a mixed train on the Tanga line.**

111. **A peaceful scene on the shores of Lake Victoria with Tanzanian 2-8-2 No. 2609 simmering beside the *MV Victoria* docked at Mwanza.**

112. **No. 2107, one of EAR's only 4-8-2 class, storms through Tanzania on a freight run west of Tabora.**

111

112

113. **Tabora shed retained the roundhouse style of the original German railway builders. Two generations of Tanganyikan steam power, 2-8-2 No. 2507 and 2-8-4 No. 3142 cluster round the turntable, with a third locomotive in the background.**

114. **A rare sight in East Africa, but John Allerton was on hand to record this northbound freight in the capable hands of two 31 Class 2-8-4s on the Nanyuki line, near Mount Kenya, in February 1977.**

115. **Although a small part of a panoramic scene including Lake Victoria, the EAR locomotive and carriage colours contrast clearly with the natural colours of the landscape as 2-8-4 No. 3101, 'Baganda', hauls the Butere passenger train between Kisian and Lela.**

113

114 115

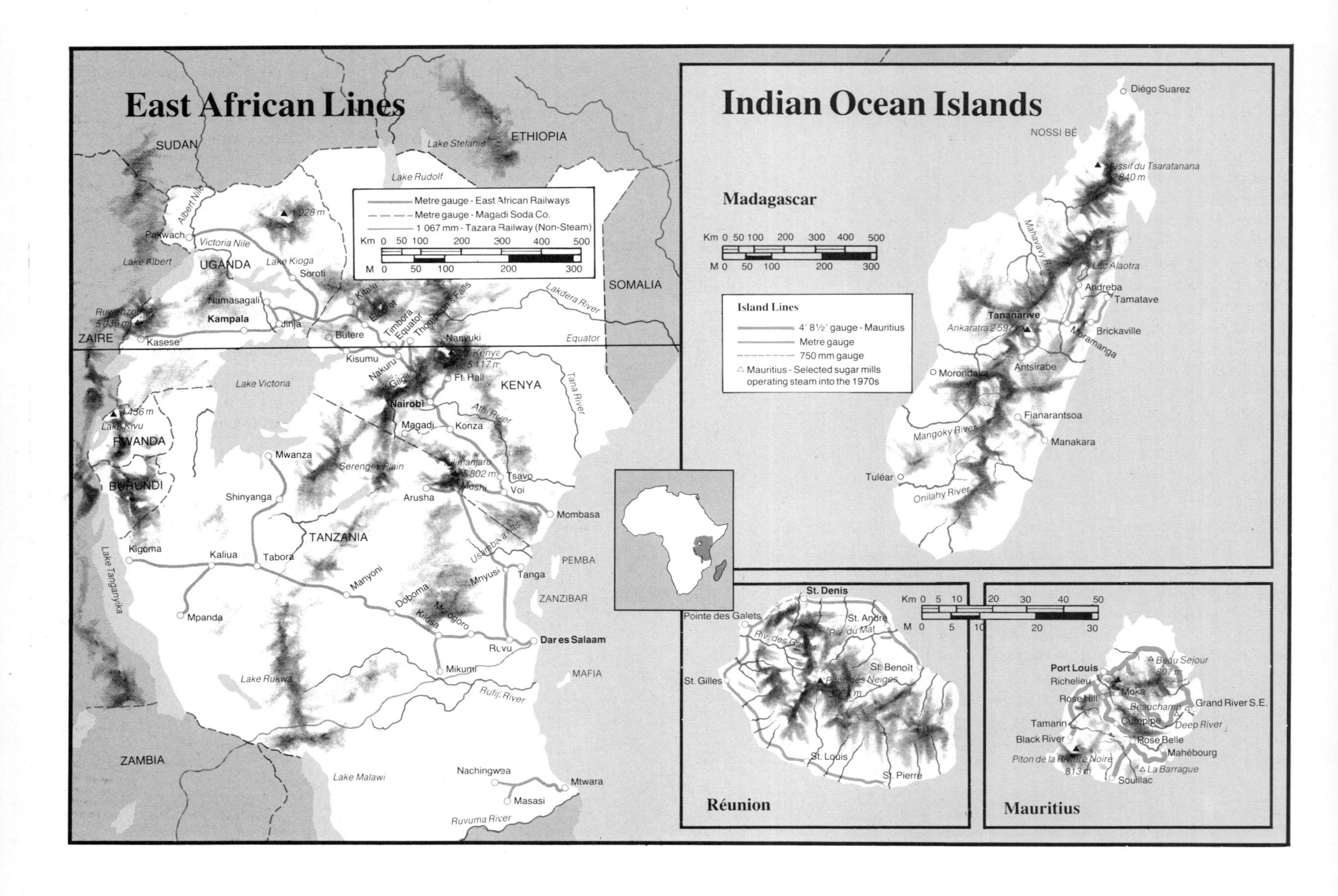

THE TANGANYIKA RAILWAY

Following Germany's defeat in World War I, Tanganyika was administered under a British mandate from 1919 and the Tanganyika Railway came into being. The Germans had pursued an extensive policy of railway destruction, so several varieties of locomotive were drafted in to replace former UB and OAEG power, some coming directly from India, others from the UR, though all were of Indian types. The A, B, and F classes used on the UR were represented as well as some Hanomag 2-6-0 class Ms from the Assam Bengal Railway. Most useful was the BESA 4-8-0, similar to those already operating in Kenya, and there were also some 4-6-0s which, with some of the 4-8-0s, were repatriated to India at the end of the war.

New construction for TR commenced with six more 4-8-0s by Beyer Peacock in 1923, while Nasmyth Wilson supplied 13 more in 1930. Surprisingly, the Beyer Peacock examples were of an improved design with superheaters and piston valves, although lacking the straight ported cylinders of the corresponding KUR class EB3. The later 4-8-0s reverted to the BESA saturated slide-valve design, probably because some lines were still unable to bear the heavier superheated version.

With the wood fuel used in Tanganyika, the small, narrow fireboxes of these 4-8-0s must have proved a limiting factor and new designs with wide fireboxes were produced, similar in concept to those already prepared under German auspices. However, these had low running plates and similar British colonial detailing. Eleven 2-8-2s built in 1925-26 gave good service, but there must have been some reservations as to the suitability of leading pony trucks for main line work, for between 1928-30 the next eight engines, although of the same basic dimensions, were built as 4-8-2s. Both designs were by Vulcan Foundry, but for some reason the 4-8-2 version suffered steaming troubles, which continued to plague them to the end.

For the heaviest shunting, TR bought four 2-6-2Ts, identical to the KUR engines, from Vulcan Foundry in 1930. Light shunting was undertaken by eight, geared high-pressure 0-4-0Ts of the well-known Sentinel design, and built between 1929 and 1931. In 1930 for really heavy work, TR took delivery of three 4-8-2+2-8-4 Garratts from Beyer Peacock, to the same basic design as the KUR classes EC to EC2, but with tall, narrow front-tanks. The depression years were particularly hard on Tanganyika, and after restocking the system with these locomotives, nothing else was acquired until World War II, when four of the WD light Garratts, identical to the KUR class EC5, were put to work. Four third-hand 4-6-0 tender engines – the only engines of this type in East Africa – were acquired in 1947 from an Egyptian site near Suez; they were of the wide firebox Ajmer design built for the BBCI railway in India.

In the immediate post-war days TR managed to negotiate the purchase of six more 2-8-2s, generally to the 1925 Vulcan design, as well as two 2-6-2Ts. These were much-modernised versions, with superheaters and piston valves, of the existing 2-6-2T. Some second-hand wartime American 2-8-2s were bought from Malaya, where they were considered greatly inferior to the splendid, locally-designed three-cylinder Pacifics. Further MacArthur 2-8-2s were bought for Tanganyika after the formation of EAR, and eventually 17 were in operation.

EAST AFRICAN RAILWAYS

The East African Railways and Harbours administration, or EAR, was formed in 1948 when the Kenya-Uganda and Tanganyika Railways were amalgamated to form for the three countries of British East Africa a common rail system. The only major point of difference was in the braking systems – Westinghouse on KUR, vacuum on the Central line and dual fitted on the engines of the Tanga line, or former Usambarabahn.

During the EAR regime, three new railway lines were completed. First of these was the Southern Province railway in Tanganyika, intended to carry export produce from the giant, but disastrous, groundnut scheme to the little port of Mtwara, using several small saturated 4-8-0s. The groundnut scheme failed and there was so little alternative traffic that the line eventually closed.

Far more ambitious was the line from Mnyusi, on the Tanga line, to Ruva on the main Central line from Dar es Salaam. This 187 km of track completed in 1963, for the first time linked every line in East Africa. Unfortunately, Tanzania's sagging economy has prevented full use being made of the line. The third was the Western Uganda Extension railway.

Locomotives

From its inception in 1948 to the delivery of the last new steam locomotive in 1955, EAR vigorously expanded its locomotive fleet with new engines of modern design, some of quite exceptional power. At an early stage, the stock was renumbered into a decimal system where, for example, class 24 was numbered 2401 upwards, class 25 was 2501 and so on, the classes being grouped according to locomotive characteristics. Thus classes 10 to 12 were tank engines, 20 to 28 were tender engines, and 50 to 58, Garratts. Full details of renumbering and reclassification have been recorded by Ramear. Much of this progress stemmed from the efforts of the Chief Mechanical Engineer, W.E. Bulman, whose previous experience in Canada had encouraged him to 'think big'. Of the classes first delivered, several were already on order from the previous KUR and TR organizations, while others were second-hand from other systems, and not true EAR designs. In this category were the class 12 2-6-2Ts, ordered by TR, but delivered after amalgamation, the later 26 class 2-8-2s of TR design, and further second-hand engines of classes 27 (USA 2-8-2s from Malaya) and 55 (WD-type Garratts from Burma). There were also the class 58 Garratts already described under KUR class EC3, and which were the only locomotives adorned with the full initials EAR & H, the others being identified by the more relevant EAR. Allowing time for specification, design, ordering and construction, it was 1951 before any new EAR types were built but, from then and until 1955, six new classes were introduced – for shunting, main line, and branch service.

One of the earliest needs was for a powerful shunting engine to cope with increased loads. The only shunting engines were the small 2-6-2Ts and 2-6-4Ts, supplemented with whatever older 4-8-0s could be spared from branch working. The main requirements were for the yards at Nairobi and Mombasa, and to a lesser extent at Nakuru and 18 of the standard industrial 4-8-2Ts produced in large numbers for South Africa by North British, were built for EAR in 1953.

The basic design of these engines dated from the turn of the century, when more than 100 4-10-2Ts were built for the Natal Government Railway. As these were removed from main line work, they were converted mainly to 4-8-2Ts, many being sold off to the mines. As they wore out in mine service, North British conceived a modernized version, which included piston valves, Walschaerts gear and a wide firebox, but was saturated and bore marks of its long ancestry – including frames cut for easy conversion to the original 4-10-2T design. As an off-the-shelf design they were better than the existing 2-6-2Ts, but were far from solving EAR's problem. They were far too light-footed and a driver could barely open the regulator without a violent slip.

The other disadvantage of the unsuperheated 13 class was its inability to complete a shunting shift without watering. When 1308 was provided with a superheater taken from a scrapped 24 class 4-8-0, the situation was improved but not cured, so for some time the whole class trailed auxiliary tanks, some of them old Garratt tanks, on 4-wheel wagon chassis. Eventually all were rebuilt under Bulman's direction as superheated 4-8-4Ts, with enlarged side and back tanks, the latter supported on bogies taken from scrapped early Garratts of classes 50-53.

For a quarter of a century motive power policy in East Africa had been mainly to provide more and more high capacity locomotives for main lines. In a developing country the demand for capacity is faster than the obsolescence rate for motive power and, within limits, it is economical to place yesterday's outclassed, but still fairly new, locomotives on secondary services. However, as traffic expands, a situation arises where displaced main line power is too big for branch lines, and the older locomotives are wearing out, creating the need for new secondary and branch engines. In East Africa, this situation was reached about the same time as the formation of EAR, and the new shunting engines described previously illustrate one facet of the solution. For branch line working and secondary duties on main lines such as pickup freights, it became desirable to replace the older 4-8-0s with something better, and another North British standard was chosen.

This was the Nigerian River class 2-8-2, which although rather a gawky and angular machine, embodied the main ingredients of the modern steam locomotive of the mid-1950s. Twenty were built for EAR in 1951-52, and another 11 in 1955, but even these presented problems. In Nigeria, as in Malawi, they represented the most modern main-line power, and the customer was quite happy. On Kenya's bigger and busier railways they were for secondary duties and in the tender-first running which such work often entails, the tenders and trailing trucks were derailed with embarrassing frequency. Bulman made several modifications to the tender mass distribution, but the 29s were not fundamentally altered.

A need also arose for main-line engines in Tanganyika to replace the rather feeble and sluggish machines of TR – classes 21, 25 and 26 – and with the upgrading of the TR main line to take a 13-tonne axle load, the 29 class design was developed to include a four-wheeled Delta type trailing truck, making it a 2-8-4. This, combined with compensated springing, greatly improved the riding, and, with their high capacity tenders on two six-wheeled bogies, they became quite respectable locomotives, spending their entire lives in Tanganyika/Tanzania. The final development in tender engines was a redesign with smaller boiler and cylinders to produce the class 31 branch-line engine, with just over 11 tonnes axle load. In 1955, 26 of the 30 class were built for Tanganyika, and 46 of the light 31 class were built for use throughout the system.

The development history of EAR's final Garratts shows that their apparent seniority is deceptive, for it places them into an order incompatible with their class numbers. Before World War II some light Garratts were designed for Brazil, but wartime restrictions deferred their construction. During the war, the design was found to be ideal where heavier power was needed for lightly built lines, and the design was revamped for use in India, Burma and East Africa. These, on EAR, became class 55, and a post-war batch destined for Burma became class 56. The success and general utility of this class resulted in a further order, and as these differed only in increased height to suit the new EAR loading gauge, they were also classed 56. However, it was eventually decided that the new dimensions needed a new classification, and as the 59 class were already on order, the new batch became class 60, 29 of which were built in 1953-54. As the most modern branch-line Garratt on the system, they were popular and saw service in all three territories. Recent reports indicate the last active engines may still be found in Uganda.

EAR's final consummated steam design was the magnificent 59 class Garratt, a 4-8-2 + 2-8-4, its dimensions surpassed only by the solitary Russian broad gauge Garratt, R-01 class of 1932, also built by Beyer Peacock. But the Russian engine was a one-off job, never repeated. By comparison, the EAR design had an initial order of nine engines, and was eventually expanded to 34 machines, the largest and most powerful steam locomotives both on metre gauge and in the southern hemisphere.

It was while the 59s were being commissioned that co-author 'Dusty' Durrant joined EAR as a junior draughtsman. He had trained at Swindon, where in the early years of the century, G.J. Churchward had produced his brilliant designs, and where at the time S.O. Ell was doing important work on locomotive testing and performance.

At the time, Willie Bulman was looking for experienced design

116. **During building of the Uganda Railway, construction work at the Tsavo River was halted for several weeks as man-eating lions terrorized the encampment, abducting and devouring the workers with terrifying regularity. Eighty years later, no sign of the former carnage remains as a 252-tonne 59 Class Garratt rumbles across the now peaceful river, although hippo footprints in the mud show that big game is still around.**

117. **Despite its proximity to the equator, Mount Kilimanjaro remains snow-capped throughout the year, forming a splendid backdrop on a clear day, for a modern lightweight Garratt, No. 6012, sitting on a train in Moshi station. Its gleaming condition displays the crew's pride in their articulated steam steed.**

118. **Doubleheading of Garratt and unarticulated locomotives was a fairly common, if not regular, operational feature on the Nanyuki line. Here 55 and 31 Class engines team up on a train composed largely of oil tankers.**

draughtsmen to work in Nairobi. 'Dusty' responded to a circular from the Crown Agents and was delighted eventually to be accepted. He recalls: 'The situation in 1955 was fluid and uncertain. There was a huge backlog of traffic in Mombasa, bound for upcountry destinations, and EAR seemed unable to cope. Bulman was the brain behind the 59 class, by then only just coming into service, and he was one of the old generation CMEs who could walk around his drawing office and comment with experience on almost any detail under way. When in a good mood, he would regale us with his experiences in the Canadian Rockies, where double-headed and banked 2-10-2s and 2-10-4s clawed over the Great Divide with never an anxious moment on steaming capacity.

'Bulman was in Gorton when the 59 class was being built, and despite Beyer Peacock's arguments on flanging blocks and press capacities, insisted that the locomotive should have a boiler of 7′ 6″ (2 286 mm) diameter, rather than of 7′ 0″ (2 134 mm). He was not going to have a big engine with a small boiler. To one who had been brought up on a railway where engines had been built on more than double EAR's rail gauge, yet whose boiler diameter had never exceeded 6′ 0″ (1 829 mm) this was big thinking, indeed, and the Great Western's most powerful locomotive, the 'King' class with 40 000 lb (18 000 kg), paled into insignificance by comparison with EAR's class 59, with its tractive effort of 37 750 kg (83 350 lb).'

Thirty-four class 59s were built by Beyer Peacock in 1955, and the last placed in service in 1956. Their impact upon the motive power can be judged by the fact that 86 Garratts of classes 50 to 58 totalled 3,98 million lb of tractive effort, while the 59 class added 2,83 million lb – a 70% increase.

'The 59s had been built as a result of the backlog of traffic which had built up and could not be moved due to inadequate railway capacity,' Durrant continues. 'It was foreseen that even the 59 class could not cope with future requirements, and Bulman set about designing an engine of even greater power that would stagger the imagination of those who think that metre gauge has capacity limitations.

'The basis of the proposed 61 class was a very much enlarged 4-8-2 + 2-8-4, with a 26-tonne axle load and tractive effort of 115 000 lb (52 000 kg), later expanded to a double 4-8-4 with 27-tonne axle load. Empirical methods were used to determine track stresses with these enormous axle loads and, as I had seen several 12-coupled types working successfully in Europe, on lines severe in both gradient and curvature, I worked out and suggested to Bulman a 4-12-2 + 2-12-4 Garratt with the same axle load as a 59, yet able to exert a tractive effort of 125 000 lb (56 600 kg).'

However, the expected extra traffic did not materialize, and the performance of the 59s exceeded expectations to such an extent that the traffic backlog was rapidly cleared, so that, unfortunately, the 61 class was never needed. For 24 years, from 1955 to 1979, the Nairobi-Mombasa main line was virtually monopolized by the 59 class, except where smaller engines were used on pickup freights, or a 58 class on the passenger work. The last class 59 has been retained in working order, and, hopefully, eventually there will be steam tours operated by EAR, using a selection of motive power.

At various times, wood, coal and oil have been used to fuel East Africa's steam locomotives, which have changed from one fuel to another according to price and availability. Neither coal nor oil is to be found in any of the East African territories, so that wood was the only available form of indigenous energy. Coal was obtained from Britain, India and South Africa, while residual oil of bunker 'C' grade was available fairly cheaply from the Middle East. In 1953, the diagram book showed all three forms of fuel in use, whereas by the late 1960s all locomotives burnt oil. The haulage of oil or coal up-country from Mombasa was always a fairly heavy expense and probably hastened the demise of steam.

Outside Europe, the EAR was the largest user of the Giesl ejector, a device for improving the front-end efficiency of a locomotive, reducing fuel consumption and increasing power output. Durrant had been in contact with Dr Giesl while in Swindon, had seen Giesl-fitted locomotives in Austria and, accordingly, suggested trials of the equipment to Bulman. This was received with much more enthusiasm than his suggestion for a 12-coupled Garratt, and the first installation was made to 6029 in 1957, followed, after two years' evaluation, by further examples on classes 58, 59 and 60. The results were so satisfactory that eventually all the post-war classes were Gieslized, with the exception of the 13 class shunters.

The EAR and its predecessors – serving areas of outstanding beauty teeming with game – have always been publicity-conscious. Whereas most railways were content to use black-painted engines identified by numbers only, the East Africa lines capitalized on the publicity value of named locomotives, painted in attractive liveries. Both the KUR and the TR gave their principal locomotives interesting tribal names, and TR instituted a red livery later adopted by EAR. In the mid-1950s, it was decided to greatly extend the practice of using names, at this stage carried only by classes 28, 50, 51, 52 and 57 in Kenya and classes 21 and 53 in Tanganyika. The new tender engines of classes 29, 30 and 31 were all given tribal names, in a few cases already carried by older power, and the 59 class was named after local mountains, with the altitude included as an interesting detail.

Until 1975, most of the more modern classes were still in service, but today, East Africa must sadly be classified as an area where steam is history. The railway museum in Nairobi can never recapture the atmosphere of steam's heyday when, over a sundowner of 'Tusker' beer, one could listen to a 58 class Garratt clawing its way up the Mau escarpment, the units wandering in and out of synchronization in that fascinating Garratt rhythm so familiar to those who know the beast well.

THE ISLANDS

5 SEVEN ISLAND LINES

Islands – particularly those in the tropics – fascinate most imaginations. For the lover of odd, quaint and distinctive railways, island rails possess a dual attraction and if continental African railways enchant, their counterparts on the various islands provide, perhaps, an even more exotic quality. All seven islands dealt with in this chapter operated railways at one time, although on most of them, the rumble of steel wheels on steel rails is a thing of the past.

Madagascar, now the Malagassy Republic, is the largest of the islands; has the biggest population and the longest railway system – the only one still in operation. Zanzibar, 1 200 km to the north-west, had a short railway, a mere 21 km, which enjoyed a short, 21-year life-span, disappearing from the scene in 1928. The railways of Réunion and Mauritius reflected their colonial masters' idea of what a line should look like. In Réunion the railway was built to metre gauge, typical French secondary-railway type locomotives being used, while in Mauritius standard gauge was chosen, complete with bull-head rails laid on chairs and locomotives which would have looked at home in the English countryside. Neither of these railways has survived, though on both islands light railways still have a limited use in the sugar-cane fields.

Off the west coast, the numerous islands are quite small and only three had railways, none of which has survived. To the north Madeira has been included, in spite of its being 700 km from the African coast and nearly 900 km from Portugal, for both have had strong links with Africa.

Neither the Canary Islands, the Cape Verde Islands, nor the Azores had railways, to our knowledge, although some push-bahns or agricultural light railways may have existed. The two other island railways were located in the Gulf of Guinea – the Spanish island of Fernando Po having had a rack railway and the Portuguese island of São Tomé a 20 km light railway which must have been one of the most picturesque of all island lines, but it was born to chug unseen by any railway enthusiast.

MADAGASCAR

As the fifth largest island in the world, Madagascar, now the Malagasy Republic, dwarfs all other African islands. This is yet another country where the French were active, although at a later period. Their presence was resisted, but the island fell to colonial rule in 1895. Soon after the conquest it was decided that the capital of Tananarive should have a rail link with the coast. Construction began at Brickaville, a few kilometres inland along a river, and progressed into the mountains – the first section of 101 km to Fanovana was opened in 1904. Another five years were to pass before the railway, now 271 km long, reached Tananarive. The completed line was an engineering masterpiece, climbing fearsome grades to an altitude of 1 430 m, passing over numerous viaducts and through 23 tunnels.

The lines which followed were easier to build. As Brickaville was found to be an inconvenient terminus, a new line was extended 97 km northwards along the coast to Tamatave where a more suitable port was developed and the railway was opened in 1913. A decade later two more lines were built. One was the 158 km extension of the main line from Tananarive southwards to Antsirabe; the other, a branch from Moramanga, 167 km northwards to Lac Alactra. Both these lines were on the highland plateaux, following mountain ranges rather than climbing them, but the next line built proved another example of mountain railway engineering. This was the 163 km line between Manakara on the coast and Fianarantsoa, which surpassed even the Brickaville-Tananarive main line for splendour, climbing to an altitude of 852 m (2 794 ft) in a distance of 42 km (26 miles). Fifty-one tunnels, with an aggregate length of 5,6 km, and high viaducts are features of this line which was opened between 1934 and 1936. After a devastating cyclone in 1934, it was decided to enclose nearly 300 m of exposed line with covered galleries for protection against rockslides.

The entire railway system has been fully dieselized since the mid-1950s, but fortunately, several enthusiasts savoured the pleasant delights of the island's railways before diesel fumes replaced the sweet smell of wood – the fuel consumed by all the steam locomotives as Madagascar has little coal and this is located in the south-west, far from the nearest rail-head.

Madagascar's first steam power was the usual small construction locomotives – a number of 13 tonne 0-6-0Ts, built by F. Weidknecht, of Paris, in 1901. Later the same firm delivered one or more 4-4-2Ts which were said to have been used on the level track out of Tamatave, but a four-coupled engine must have had limited use on such a mountainous line. It is natural that on a French-controlled railway articulated locomotives in the form of Mallets should have made their appearance. These were lovely little machines, built by Batignolles in 1906, featuring slide-valves all around and being tank-tender locomotives. The small tender attached was only for fuel, and liberal quantities of wood piled high gave these engines a distinctive appearance.

Few English-speaking people were aware of Madagascar's railway or its locomotives and a wartime visit by William F. Bolton, soon after the British

120

121

119. **An island of steam – on an otherwise largely dieselized railway. In 1954 when C.S. Small visited Madagascar the isolated Manakara-Flanarantsoa line was, apart from diesel railcars, 100% steam. At Flanarantsoa he found No. 21.946, a 1922 O&K 0-6-0T for shunting, and a main-line Mallet building up steam behind. There is no obvious shortage of fuel here.**

120. **A classic 0-4-4-0T+T Mallet, No. 32.851 – a 1916 SACM product – sits at Moramanga shed, midway on the main line from Brickaville to Tananarive, junction of the branch to Lac Alactra.**

121. **This rare Garratt, No. 101 (later 59.801) was built by St. Léonard in 1925 – designed from the start for wood burning – as the spark-arresting chimney and slatted high-sided fuel bunker indicates. It is surprising that this design did not have a greater impact on the railway, for by comparison with the standard 0-4-4-0T+T Mallets it had nearly twice the tractive effort – its adhesion mass being 59 tonnes to the Mallet's 32 tonnes. Perhaps these two Garratts were too powerful for their time – a typically shortsighted view encouraged by exponents of diesel traction in later years.**

occupied the island in 1942, is illuminating. He wrote of his first visit to Tamatave railway station:

'I entered the platform on my first visit just as a train was due and had no idea what to expect. It was a queer sensation. Standing on the platform of a railway whose existence I had not known of a few weeks before, some 6 000 miles from home – and waiting for a train to appear without having the slightest idea of what it might be like.

'I have had a certain knowledge beforehand of all the countries I have visited; I knew the types and size of their locos, etc., but of the Madagascan line I knew nothing. In steamed the train and drew to a standstill some distance from the stops. The engine at once drew my attention. It was like nothing I had ever seen before on five continents. A short stubby articulated side tank of the 0-4-4-0 wheel formation having a tender so small that I could hardly believe my eyes. A tiny four wheel affair carrying fuel only and having the wire-mesh compartment above its sides. I was almost afraid to step on it from the footplate in case it should tip up with my weight. This was No. 68 of the TCE.'

No. 68 was a 1925 SACM-built locomotive of the Tananarive-Côte Est section of the Chemin de Fer Tananarive à Tamatave Railway and one of more than 50 of these 0-4-4-0 tank-tender engines built between 1906 and 1930 by SACM, Batignolles, and even Baldwin. The first Mallets were unsuperheated with slide valves, but on the later engines the high pressure cylinders had piston valves and the boilers were superheated. A number of 2-4-4-0Ts were bought from the Swiss Rhätische Bahn when they electrified during the 1920s and later some Tunisian (Bône-Guelma Ry) 0-4-4-0Ts supplemented CFM's own Mallets, making these articulated engines the mainstay of the railway. Two Garratts came to the island in the mid-1920s but, as in Ethiopia, their impact was limited though they lasted until the end of steam. These dia-

122

125

122. No. 42.104 was formerly Bône-Guelma Railway No. 479, which operated in the desert north of the Maghreb. Constructed by Batignolles in 1906, it was seen out of service at Tananarive in 1954.
123. Another second-hand locomotive, No. 40.824, was one of eight 2-4-4-0Ts built by SLM for the Rhätische Bahn in 1902, three of which were purchased by the CFM in 1922. With a mass of 45 tonnes, these engines were the largest Mallets on the island, but were early casualties when the main line was dieselized in the early 1950s.
124. From the trenches of the Western Front during World War I, to the rows of stored engines at Moramanga, this Baldwin 0-6-2T has come a long way. It was originally built as a 2-6-2T.
125. This eight-coupled Corpet-Louvet engine derelict at Tananarive is also second-hand, formerly C.F. Department Ille et Vilaine. Other steam locomotives from this railway found their way to nearby Réunion.
126. Ugainly – but at least in steam. This 1922 Jung was active at Moramanga in 1954.
127. An early Madagascar engine, No. 13.981, one of two 1901 Weidnecht 0-6-0Ts.

mond-stacked 2-6-2+2-6-2s were built by St. Léonard and incorporated such modern features as superheating and piston valves but, as was often the case, few builders other than Beyer Peacock were able to design a really successful Garratt and the order was not repeated.

Lesser locomotives included some 0-4-0Ts built by Borsig and Jung in the late 1920s and early 1930s, a few Baldwin 0-6-2Ts built for the French War Department in 1916 and acquired second-hand, a 1922 O & K 0-6-0T, and two 1931 Corpet Louvet 0-8-0Ts – the only 8-coupled engines on the roster. At its peak, nearly 100 steam locomotives were on the roster, but these had been reduced to about 30 by 1954, shortly before complete dieselization, and apparently not a single engine has been preserved.

A number of smaller isolated railways operated on the island at one time or another. In the north, at Diego Suarez, a 600 mm line served a French naval base, running up the main street of town. A visitor in 1953 found a Decauville 0-4-0T and a heap of track piled up in a side street. The nearby island of Nossi-Bé is blessed with a high rainfall and rich soil, making it a centre for Madagascar's sugar industry. Naturally, a plantation railway existed, and when C.S. Small visited the island in 1953 he found that it was metre gauge and although the line was not in operation, three H.K. Porter 2-6-0s of about 1919 vintage were stored. Another isolated railway was an 'industrial' line, running from Morondava on the Moçambique Channel, inland to Four à Chaux, but its purpose is unknown to us. Better known is the metre-gauge railway which ran from Scalary to Vohitsara serving the low-grade coalfield in the area. This railway is long abandoned and records of its locomotives remain hidden.

MAURITIUS and RÉUNION

Far removed from the African mainland, Mauritius and Réunion, in spite of being only 150 km apart, differ greatly from each other in topography though both are volcanic in origin. Railway development reflected their differences. In Réunion a central mountain range reaches an altitude of more than 3 000 m, while in Mauritius the peaks are not higher than 850 m, but are placed more randomly around the island – allowing railways to cross its centre rather than encircle it. Sugar-cane, the mainstay of both islands' agriculture, led to the large-scale introduction of sugar tramways which are today the last vestiges of rail transport on both islands.

Their climates are tempered by the same trade winds which often stir up severe gales and even cyclones. On Mauritius in the 1890s a gust of wind blew almost an entire train off a high bridge.

Passengers on each of these islands travelled through very different scenery, sugar-cane plantations notwithstanding.

Mauritius

Smaller of the islands but with a much larger population, Mauritius had the longer railway system – 225 route km of track. Opened in 1863, it was one of Africa's earliest railways, coming only nine years after the first railway in Egypt and three years after Natal's. In a continent better known for its narrow-gauge railways it is interesting that – as in Egypt and South Africa, where the first railways were standard-gauge – Mauritius also had this wide gauge. Of all the islands it was the only one to have such an extravagant track width.

Development was rapid and by 1880 all but two of the branches were completed to give more than 166 km of line and some 30 steam engines were in operation.

Since the longest run on the island was about 50 km, tank locomotives became standard, although the first eight (Nos. 1 to 8) were 0-4-2 tender engines built by Sharp Stewart between 1863 and 1865. The first four of these had 1 524 mm (5′) coupled-wheels, and the second four 1 676 mm (5′ 6″) coupled-wheels – largest of any locomotives to operate on the island. These engines were followed by seven double-frame 0-6-0Ts, numbered 9 to 15, and constructed by Sharp Stewart in 1865. Nine larger 0-8-0 saddle-tank engines (Nos. 16-25) came from Sharp Stewart and Robert Stephenson between 1866 and 1869. These eight-coupled inside-cylinder engines were very similar to a batch built for the Great Indian Peninsular Railway, which used them up the Western Ghats section and, no doubt, the Mauritians thought the design suitable for their own mountain railway. This was the Midland Main Line, where grades varying from 1-in-30 to 1-in-25 brought the line to a maximum altitude of 552 m (1 812 ft) at Forest Side, 27 km from Port Louis. Heavy seasonal sugar-cane traffic required as many as three engines per train – until the advent of articulated power in the 1920s.

Before the delivery of articulated locomotives however, a variety of inside- and outside-cylinder engines, some with double frames, with wheel arrangements of 0-4-0ST, 0-6-0T, 0-6-2T, 0-8-0T, 2-6-2T and 2-8-2T arrived which, with the Garratts, brought the all-time steam roster to 81 locomotives. One unusual locomotive, and the only one acquired second-hand, was a former Great Central Railway 0-6-2T, built in 1885, and transported to Mauritius in 1920.

The largest Mauritian locomotives were three 2-8-0+0-8-2 Garratts, constructed in 1927 and intended for heavy freight haulage. These engines weighed 154 tonnes, had a 16 tonne axle load and exerted a tractive force of 28 725 kg (63 340 lbs) at 85% boiler pressure.

Unfortunately, these engines arrived shortly before the onset of the Depression and Mauritius never fully recovered from this economic downturn so that the Garratts' full potential was rarely utilized. As traffic fell off it was more economic to use the smaller engines and toward the end the Garratts were seldom used.

The last new steam locomotives were bought in 1930, at the onset of the Great Depression. World War II restricted the flow of sugar exports and in the immediate post-war years road transport hit the railway hard, leading to ever-increasing losses. These resulted at first in the closure of branch lines; then an end to passenger services in 1956; and a general running-down of plant until 1963 when only 109 km of track remained of which only 41,5 km was still in use. The following year the system closed.

In its last year 37 steam and two diesel locomotives were on the railway's books, only a fraction seeing regular service. They included No. 13, built in 1864, which reached its centenary as the railway closed. Today only occasional culverts and embankments remain. None of the standard-gauge steam locomotives was preserved but a fine impression of the railway appeared in *Island of the Swan* where Michael Malim described his experience at Curepipe Station in the late 1940s:

'The train came in fussily, drawn by a little black engine with the island's arms emblazoned hugely on each tank. There was a frenzied rush for seats. The coaches were mere boxes, khaki-coloured, each on four wheels, with a large 1, 2 or 3 painted on each door. One would not have been at all surprised to see a sheep knitting at the window of any of them. The class distinctions were marked: 1 meant a leather seat, 2 a wicker one, 3 mere wood. Broad wooden sunshades jutted out over each window.

'An enchanting train. As we took it in, the assault swept past. The ladies, one notices, got, and evidently expected, no quarter. We missed our chance and looked, I suppose, rather helpless. But a splendid person came to our rescue, dressed all in blue with a green-banded cap which had a bobble on top of it worthy of a French admiral. He presented himself with a smile, bowed, produced a great key, "cleared all sea-roads" as it were, and led us down the train till he came to a door marked 1 which he opened with a fine flourish. He held the crowd at bay with the majesty of a Mussolini. As soon as we were inside he locked the door again.

'The gesture itself was enough to endear the Mauritian Government Railways to us. That carriage was fragrant with the mustiness of an old berlin [hat]. We sat down in it gingerly, rather awed. We were distinctly vulgar

128. **These ancient wooden 'double-deckers' are being pulled by an appropriately antique-looking Kitson 0-6-2T.**

129 130

129. **Réunion possessed many locomotives, though few photographs of them have survived. Fortunately engine No. 8, one of a group of nine 1878 Schneider engines – the first on the island – has been preserved at St Denis.**

130. **By the early 1960s the writing was on the wall for the public railways of Mauritius, though those engines still operating were clean. At Port Louis an 0-8-0T, No. 74 built by Kitson in 1930, and No. 15, an 1865 Sharp-Stewart 0-6-0T, simmer between turns, which were already decreasing. No. 15 was one of seven double-frame 0-6-0Ts – the first tank locomotive on the island. She originally carried the name 'Firefly' for in earlier days all Mauritius Government Railway locomotives had such titles as 'Vesuvius', 'Rocket', 'Cyclops', 'Dodo', 'Lord of the Isles' and 'Flacq'.**

131. **Kitson and Vulcan built six large 2-6-2Ts in 1920-21, and No. 22 was active hauling freight from Port Louis during the last years of the Mauritius standard-gauge railway.**

132. **The last active steam engine in Mauritius is this 800 mm-gauge 0-6-0T, looking lonely on the Beau Champ Sugar line, at Deep River. Before World War I, nearly 60 sugar factories operated hundreds of kilometres of track on the island and a fascinating variety of locomotives must have existed. Unfortunately few locomotive historians have explored the records, and the story of these railways remains to be told.**

133. **Academically interesting, and attractive in a fresh coat of paint, this O&K 0-6-0T would be happier in steam.**

134. **With remarkably primitive wheels, this old 0-4-2T, named 'Hariette', was built by the little-known firm of Fletcher-Jennings.**

131

132

133

134

135. **On the Midland main line, Garratt No. 62 sits with a heavy goods train. These massive machines could haul a 350-tonne train on the 1-in-25 climb to Rose Hill and, according to an official statement, 'obviated the employment of working double and even treble headed trains, and also all night working'.**

anachronisms. The engine piped a frail 'toot' and tugged us impatiently out of the station. We clacketed briskly along from stop to stop. They had curiously assorted names: Floréal, Vacoas, Phoenix, Quatre Bornes, Beau Bassin, Rose Hill, Richelieu, and Pailles. As we ran down the hill towards the sea the guard came clambering along the running board and put his head in through the window to ask for our tickets. And ghosts came with us all the way, genteel ghosts but warm. There hung still about that ancient coach something of the thrill its first passengers must have known. The craftsmanship of its eighty-year-old fittings made one suddenly homesick for the standards of a prouder and more generous age.'

That age included a myriad of narrow-guage sugar-cane railways, even one government-owned 760 mm (2′6″) gauge line which operated three O & K locomotives. This 16 km line opened in 1904 and closed soon after the end of World War II.

The cane railways originated in the late 19th century – the result of a cattle plague which decimated the island's draught animals and forced the introduction of mechanical transport. These tiny railways lasted longer than the standard gauge and some are still in use. A great variety of gauges were to be found, including 600 mm, 750 mm, 760 mm, 800 mm, 920 mm and 940 mm. By 1978 only one cane railway retained working steam power; the Deep River Beauchamp Sugar Factory, where three 920 mm-gauge locomotives were to be found. The engines were all different and included 0-4-2T No. 204 'Regina', built by the Lowca Engineering Co., Whitehaven, U.K., which was on display. Out of service but intact was 0-4-2T No. 190, 'Harriette', built by Fletcher Jennings in 1883, and finally 0-6-0T 'Sir William', built by O & K. This engine was featured on a Mauritian postage stamp issued in February 1979, and a similar but 800 mm-gauge engine was retained on standby for a fleet of nondescript diesels.

Réunion

Nearby Réunion's railway boasted a tunnel 11 km long – one of the world's longest. There were also numerous bridges, including a 375 m structure spanning the Rivière des Galets and another 480 m long across the Rivière du Mat.

Réunion, like Mauritius, was discovered by the Portuguese explorer Mascarenhas in the early 16th century but – unlike its sister which was occupied and named by the Dutch, abandoned to the French and finally taken over by the British during the Napoleonic wars – Réunion has always remained French, except for a short period of British occupation from 1810 to 1815. Today it is one of the last French colonies, administered as a *département*.

In the latter part of the 19th century the crescendo of railway building in Europe and America encouraged the people of Réunion's capital, St. Denis, to have a railway of their own. For years St. Denis had been the main port for the island, but it was badly situated and, as the island possessed no natural harbours, a more suitable location for an artificial port was sought. A site was found and the engineers ingeniously designed a small port 3 km north of the mouth of the Rivière des Galets, naming it Pointe des Galets.

However, a huge dividing massif of basalt separated St. Denis and the harbour. Only by driving a long tunnel through the rock could a rail link be achieved. This resulted in one of the great railway engineering feats of all time – rather out of place on a light metre-gauge railway which in its hey-day was only 126 km long. The line circumvented 60% of the coastline, running from St. Benoit to St. Pierre.

William Dudley Oliver visited the island in the early 1890s and described his arrival at Pointe des Galets and the railway in his book *Crags and Craters:*

'The train left at 9.40 a.m. and reached St. Denis at 11 a.m. The distance from the port to St. Denis being 12 miles [20 km], it will be seen that the speed is not unduly rapid. The railway is really little more than a tramway. It is of metre gauge, metals and rolling-stock are of the lightest description.

'Between the port and St. Denis there are two intermediate stations – Possession and Grande Chaloupe. The most remarkable feature of this piece of railway is the tunnel which begins soon after leaving Possession, and with the exception of two short breaks continues nearly to St. Denis. In this nine and a half miles [16 km] there are six and a half miles [11 km] of tunnel. The mountains at this point fall abruptly to the sea and the only way of taking a railway along this part of the coast was by tunnelling through the rock. There are galleries at intervals running out to the face of the cliff, which were made for the removal of the excavated rock. Altogether, one has about three quarters of an hour in the tunnel and in hot weather it is stifling.'

Oliver went on to describe the railway's locomotives as could only a Victorian English gentleman, used to railways 'at home':

'The locomotives are curious constructions, something after the style of contractors' engines, but of more clumsy build. They are about the most hideous things I ever beheld.'

At the time the railway's locomotives were unremarkable – all being 0-6-0Ts. The first nine were built by Schneider between 1878 and 1883, and were probably the only engines running at the time of Oliver's visit.

In 1900 Weidknecht built three 0-6-0Ts, numbered 10-12, and these were followed by seven Decauville engines – Nos. 13 to 19 – built between 1913 and 1920. Though uncertain, it appears that the next engines, numbered 20 to 25 and built by Soc. Alsacienne in 1883, were second-hand, as were 26 and 27. These latter were built in 1901, but acquired only about 1941. Until then all CFR engines had been six-coupled tanks, but now a new type made its appearance.

The new engines were typical 'secondary-railway' Mallets, and at least 13 0-4-4-0Ts have been recorded. Most are presumed to have been second-hand for, though their builders are known, their sources are not. Finally, in the late 1940s, 11 Corpet-Louvet tank engines, 0-6-0Ts and 0-6-2Ts, came to Réunion from the Chemins de Fer d'Ile-et-Vilaine, seemingly the last steamers acquired. A roster which appeared in *Rail et Route* of September 1954, lists only 46 engines, though the total steam roster certainly exceeded 50.

Statistics published in the 1955 *Directory of Railway Officials & Yearbook* indicate that 38 steam and four diesels were on the roster. When the line closed in 1963 only 19 steam and four diesels remained. However, these figures are suspect. When C.S. Small visited the island in 1954 all he could find were the derelict hulks of several engines in the harbour area. The railway seemed on its last legs, with a skeleton service for passengers, and freight traffic limited to occasional trains from the harbour when cyclones prevented road traffic from using the new mountain highway. Reportedly, the section between the harbour and St. Denis is still used when such conditions prevail.

ZANZIBAR

Zanzibar, like much of Africa, has suffered political upheavals in recent years but, with the right wind, the scent of cloves still reaches the mainland. It is said that Zanzibar's story has been 'written by the winds', a reference to its historic sea-trade links with India and the Arab countries which have made it culturally more Asian and Arab than African. Since its federation with Tanganyika to form Tanzania, this is changing. The island lies some 30 km off the coast and, unlike many of its counterparts, is not volcanic, having a maximum altitude of only 100 m. This presented few obstacles to rail construction; similarly there were few to road building which led to the railway's early demise.

The Sultan imported Zanzibar's first locomotive, though whether this was a mere toy or for a plantation is unknown. In 1880 Bagnall delivered this tiny, 600 mm-gauge engine named 'Sultanee', which seems to have disappeared into history. Apparently American entrepreneurs instigated Zanzibar's next railway project when, in 1906, a concession was granted to construct an electric power station and road-side tramways. The following year a 914 mm (3′ 0″) gauge railway opened, running 11 km from the old Customs House in Zanzibar town parallel to the coast to Bububu. The impact of such a short railway is not known – except that the residents of the town were not pleased with the noisy, dirty train which rumbled through the narrow streets, showering everything with dust and cinders.

In 1911, the government acquired the line and bought additional locomotives and equipment, perhaps in anticipation of a proposed, but never built, extension northwards to Mkokotoni. However, in 1921 an 11 km southward branch to a quarry at Chuckwani was built to transport stone for the construction of the new wharf, a project eventually completed in the late 1920s. The finishing of this project coincided with the abandonment of the entire railway, already suffering badly from road competition. Possibly the attitude of the railway's operating department contributed to its demise, as contemporary writers have described its operation as 'eccentric'.

The few locomotives were not without interest, though no complete documentation has come to light. It is believed that the first engines were American Porter-built 0-4-4Ts, similar to two supplied later, in 1921. The only other known engine was a dainty Bagnall 0-4-2T, supplied in 1919. One assumes that they were all scrapped when the railway closed, but it is not impossible that one or two were used later on agricultural railways. Light railways operating steam locomotives have been known to burn anything from coconut kernels to bagasse, even manure, and the imagination runs wild with the thought that these of Zanzibar might have burned cloves.

136. Built for the tropics. Everything one might expect of a railway in Zanzibar comes alive with this Porter 0-4-4T on a train of 'air-conditioned' carriages. Note the unique chimney extension, obviously designed to deflect any stray cinders which might rain down on passengers.

MADEIRA and FERNANDO PO

Madeira, like Réunion, has a beautiful mountainous interior which has been a successful deterrent to railway construction and it is no wonder then, that its only known railway should have been a rack line. In the early 1920s a British firm was awarded a contract to build a harbour at Funchal and construct 22 km of light railway westwards along the coast. What became of this project is uncertain.

Opened in 1892, the rack railway climbed the mountainside behind Funchal, the island's capital. Sadly, this railway was seldom noticed by the mass of tourists who visited the island mainly to ride the 'tea-trays' – a type of toboggan or sledge – the standard form of downhill transport through the streets of town.

The 4,8 km line was built on a 1-in-6 gradient using the Riggenbach rack system. Passengers sat in attractive open-sided carriages which were pushed uphill by either of two rack engine types. The first was an 0-4-0T true Riggenbach engine where the drive from the cylinders went directly to the rack pinion, while the flanged wheels were only for carrying. The second engine was an 0-4-2T with cylinders high astride the boiler driving the carrying wheels, whose axles were fitted with a toothed-wheel to mesh with the rack line. Both were true mountain rack engines having tilted boilers which were

137. **Though this little-known rack railway on Madeira no longer exists, the cobbled track which flanked it remains a popular spot for local youths who use tin trays as 'sledges' to descend the slope. The sophisticated 'sledge' in the foreground of this historic photograph employed two islanders as 'brakes'.**

138. **Dr Ransome-Wallis photographed L-5, an SLM 0-4-2T rack-locomotive at Funchal in 1936.**

139. **A train for all seasons on an island of fantasy. During the heyday of rail transport in São Tomé, numerous plantation railways penetrated the island's mountainous interior. A few are thought to exist today, although steam power – and passenger trains – of this sort are probably history.**

140. **A very substantial station, on a not so substantial island railway. This Maffei 0-8-0T starts a train, two carriages of which eventually found their way to Moçambique.**

138

139

140

level on the climb. Dr Ransome-Wallis was one of those fortunate to see this railway and he wrote in *On Railways at Home and Abroad:*

'I know of many seasoned travellers who claim to know Madeira well, yet few know even of the existence of this interesting little railway, and fewer still have travelled on it.'

Fernando Po (now Macias Nguerra) also had a rack railway although we know little of it. This island is slightly larger than Madeira, but with a population of only 40 000 (Madeira has over 300 000) could hardly be expected to support a major railway. This railway was opened in 1912, built to a gauge of only 600 mm, and was a mere 1,5 km in length. The rack section was of the Abt type and was just 400 m long, the gradient being 82 mm/metre – not particularly steep. Two small rack engines were supplied at the line's opening but what has become of these engines or the line is not known.

SÃO TOMÉ

Further into the Gulf of Guinea, the island of São Tomé possessed one government public railway and several plantation lines. While visiting the João Belo railway in Moçambique, rail-photographer Chris Butcher noticed that the axle-box covers on some of its passenger carriages carried the initials CFST. Inquiries revealed that this stood for the Caminho de Ferro São Tomé, the carriages apparently having been obtained when the railway closed.

Precisely when this occurred is uncertain. But in *Die Kolonialbahnen* (Baltzer) details indicate that a concession was granted in 1906 for a 40-km line running from the port of Anna de Chaves to San Sebastian. However, it is believed that only 20 km of line was opened – in 1913. Three years earlier, Maffei built two 0-8-0Ts for the railway, but what other locomotives there were remains a mystery.

Known as the 'Chocolate Island', São Tomé had numerous coffee, cacao and palm-oil plantations. The cacao plant flourishes on densely-forested mountain slopes at altitudes between 500 and 800 m. Some of the plantation railways served these areas and were truly 'mountain railways'.

A *National Geographic* correspondent visited the island soon after World War II and in an article dated May 1946 he said:

'One morning we donned cork helmets and warm overcoats and set out by rail for a trip to the peaks, which rise to more than 6 500 ft. Mules pulled us, since locomotives were too heavy to make the grade.'

One wonders what he meant by this last statement – perhaps he was referring to the weight of the engine being too heavy for a section of the line. If not, all the world's railways would surely have switched to mule power, instead of diesel and electric traction! Yes, there is nothing like a tropical island to stir the imagination.

WEST AFRICA

6 FORMER FRENCH and GERMAN COLONIES

By the late 19th century all the colonial powers had snatched their own particular portions of the extensive West African coastline with its rich – and sometimes not so rich – hinterland. Britain gobbled Sierra Leone, the Gold Coast and Nigeria; Spain snatched the Rio de Oro and the little enclave of Rio Muni, or Spanish Guinea; Portugal claimed the Cabinda strip and a small section of Guinea; Germany had a firm grip on Togoland and Kamerun; and France – already a force in North Africa – spread her colonial fingers over Senegal, Guinea, the Côte d'Ivoire, Dahomey and the French Congo. Liberia, the slave state, was already independent.

Into this patchwork of jungle, mountain and desert, only Britain, France and Germany were to introduce railways to their colonial possessions and, apart from the British lines discussed in the following chapter, only the French rail link between the Senegalese ports of Dakar and Saint-Louis came into operation before 1900.

FRENCH WEST AFRICA

On a map of Africa in the early 1920s an enormous swathe of land is shown stretching from Algeria on the Mediterranean, down to the Gulf of Guinea, with the omnibus title of French West Africa. British and other possessions appear merely as enclaves fringing much of the coastline but intruding on little of the interior. Much of the land mass straddling the Tropic of Cancer formed the great Sahara desert, and was of little use to man or beast. Further south, near the legendary city of Timbuktu, flowed one of Africa's great waterways, the river Niger, important for both transport and irrigation, and towards which all the railways built by the French colonists reached.

For several years there was, as in the Belgian Congo, an isolated stretch of line by-passing an unnavigable section of the upper Niger.

At one stage all the various lines were under a single colonial administration, locomotives being transferred from one line to another as needed. During this period of common management the locomotive stock was renumbered in a strangely Gallic decimal system, in which an apparent class number in reality denoted the adhesion weight in tonnes. This clearly was of use to the *chef de depot* in, say, Dahomé when faced with an unfamiliar machine shipped out, perhaps, from Mali or Senegal; for, the very engine number gave a close indication of what the locomotive would pull and where it could be allowed to go.

Comparatively little is known of the French colonial locomotives as steam disappeared fairly early from these areas. Thus the operations of these railways remain conjectural. Was there, perhaps, a tortuous climb up the coastal escarpment, or through an inland mountain range, where a Garratt and a 4-6-0T struggled through the dripping jungle at the head end of a heavy train, while Golwé and Mallet articulateds shoved hard behind the *fourgon?* The locomotives available would certainly have permitted such awe-inspiring battles of steam and steel versus gradient and greenery, but whether these occurred is pure conjecture.

TOGOLAND

The Germans, despite the Kaiser's expansionist dreams, were late starters and, because of those very dreams, early finishers. Their first railway was a 750 mm-gauge line which ran all of 300 m – the length of a substantial pier stretching across the pounding surf of Lomé, in Togoland – along which cargo was moved from ships moored at the seaward end to the shore. Begun in 1902, in the following two years the pier and its line were extended a further 200 m.

Initially Togoland offered little of value to its colonizers, but railway building continued nevertheless. With a coastline of less than 75 km fronting a narrow corridor of hinterland more than 600 km long, the country was ideally suited to a single inland railway which would never be far from its borders, so the decision to push the 750 mm pier track eastward from the harbour to the tiny port of Anecho, 45 km away, may have been a bureaucratic blunder rather than a move based on economic realities.

But Teutonic efficiency soon rectified the situation. Richly fertile hills and valleys lay behind the coastal strip – lands supremely suited to the cultivation of cocoa and coffee and for these to be developed and then exploited, a railway was imperative. The Anecho track may have been a fever-induced aberration, but the metre-gauge 'Cocoa Railway', or Kakao Bahn, was a practical, down-to-earth expression of faith in the country's development potential. This 119-km line, from Lomé to Palime, was started in 1904 and opened in 1905, economics providing the spur which the uninspired pier railway had lacked.

Almost concurrently work began on the main inland railway to Atakpame, opened in 1911-13 and 170 km long; in 1907 the coastal line was converted to metre gauge, standardizing the colony's rail network.

Following the World War I defeat of the Germans in Togoland by Anglo-

French forces, in 1920 the western part of the country, bordering on Lake Volta, was incorporated into the Gold Coast. The remainder of the country, in which all the railways were concentrated, came under French control. The Togo-Eisenbahn's pre-war plan to extend the main inland line from Atakpame to Blita was carried out by its French successor, Chemins de Fer du Togo, though other German-planned extensions of the railway – almost to the northern border – were scrapped.

No records, either of the locomotives used on the pier or the original line to Anecha are available – possibly it was operated by second-hand contractors' engines. Henschel supplied three 0-4-0Ts in 1904 which were probably built to the narrow gauge, later being converted to metre. Much later, in 1909, a special long-wheelbase 0-4-0T arranged to burn coke was built by Borsig for the pier railway, but details of these earlier locomotives are vague. Orenstein & Koppel built four 0-4-0Ts in 1908-09 for the Deutsche Kolonial Eisenbahn Bau- u. Betriebs Gesellschaft (German Colonial Railway Construction and Operating Co.) and these, doubtless, were used in the building of the line to Atakpame.

Earlier, in 1905, Orenstein & Koppel supplied six 0-4-4-0T Mallets to Lenz & Co and these were probably used on the Palime line. For the Atakpame line, the same builders supplied six of the standard German colonial 2-8-0Ts in 1909 – the last locomotives supplied under the German administration. By 1914 route length of the Togo Eisenbahn totalled 327 km, and of the 18 tank engines then in operation 15 have been positively identified.

Under the post-war French administration all new locomotives were of the tender type and included four 4-6-0 passenger engines by Nasmyth Wilson. In 1930 Orenstein & Koppel supplied a further two 2-8-0s and these were followed by eight superheated 2-8-2s from the Belgian builders Haine St. Pierre. Six more 2-8-2s supplied by Corpet Louvet in 1949-50 were the last steam locomotives built for Togoland.

141. **The last operative steam in Cameroon was this delightful Orenstein & Koppel 0-8-0 'Gouverneur Ebermaier', belonging to the Cameroon Development Corporation, seen here with a special train hired by Peter Bagshawe, traversing the West African jungle across the spindly Sonne Bridge in 1977.**
142. **A pair of unusual locomotives on the Togoland Railway were also 2-8-2Ts Nos 30 and 31, built, according to the Alco catalogue, for the Togoland Military Railway, although this is not confirmed by the numberplate.**
143. **Typical of the light Mikados built for French colonial service, Togoland No. 101 poses with crew at Lomé before World War II.**
144. **Among the equipment drafted in to rehabilitate the Togoland Railway after World War I was a batch of standard Indian B.E.S.A. 4-6-0s, typified by No. 42, seen with workers at the reconstructed Togblékope Bridge.**

145

146

147

145. A peaceful Togoland scene with the 2-8-2 No. 105 arriving on a freight at Atakpamé station, the period betrayed by the attendant motor car.
146. 'En Voiture!' A long past colonial era is delightfully portrayed in Yaoundé station as pith-helmeted staff attend to the departure of a wood-burning Cameroun 4-6-0T, No. 154.
147. Only Germany thought in terms of 0-8-0 tender engines for narrow-gauge colonial operation, perhaps as scaled down versions of the ubiquitous Prussian G8 class. Cameroun 0-8-0 No. 105 heads a freight through Douala in about 1927.

KAMERUN *(The Cameroons)*

Writing of his experiences in the Cameroons in the 1930s (by which time the railways were under French control) a visitor was to record: 'The road to Yaoundé was not yet completed. The train went there twice a week. We piled ourselves into it with all our luggage . . . The carriages for white men were clean and comfortable, those for the blacks, crowded to the roof. I could not tear my eyes from the window. The sight of the trees moved me most; flawless columns, smooth and branchless, towering to immense heights.'

The equatorial forests of southern Kamerun, the leech- and mosquito-ridden home of vipers and other animal hazards to the unwary railway builder, might have daunted most administrations but the Germans set about developing a rail system with the same slowness but more method than they had shown in Togoland. The country was considerably larger, had proportionately fewer railways, but the eventual total route kilometrage was greater.

The first railway was a 600 mm-gauge Feldbahn built by the Westafrikanische Pflanzungsgesellschaft Viktoria, from the port of Viktoria up the escarpment to Buëa, 43 km away. It was completed in 1901 and 0-6-0Ts and 0-8-0Ts of the usual German Feldbahn type hauled freight and passengers on the line for more than 50 years. Today the railway is owned and operated by the Cameroon Development Corporation.

Real railway development was in the hands of the Kamerun Eisenbahn

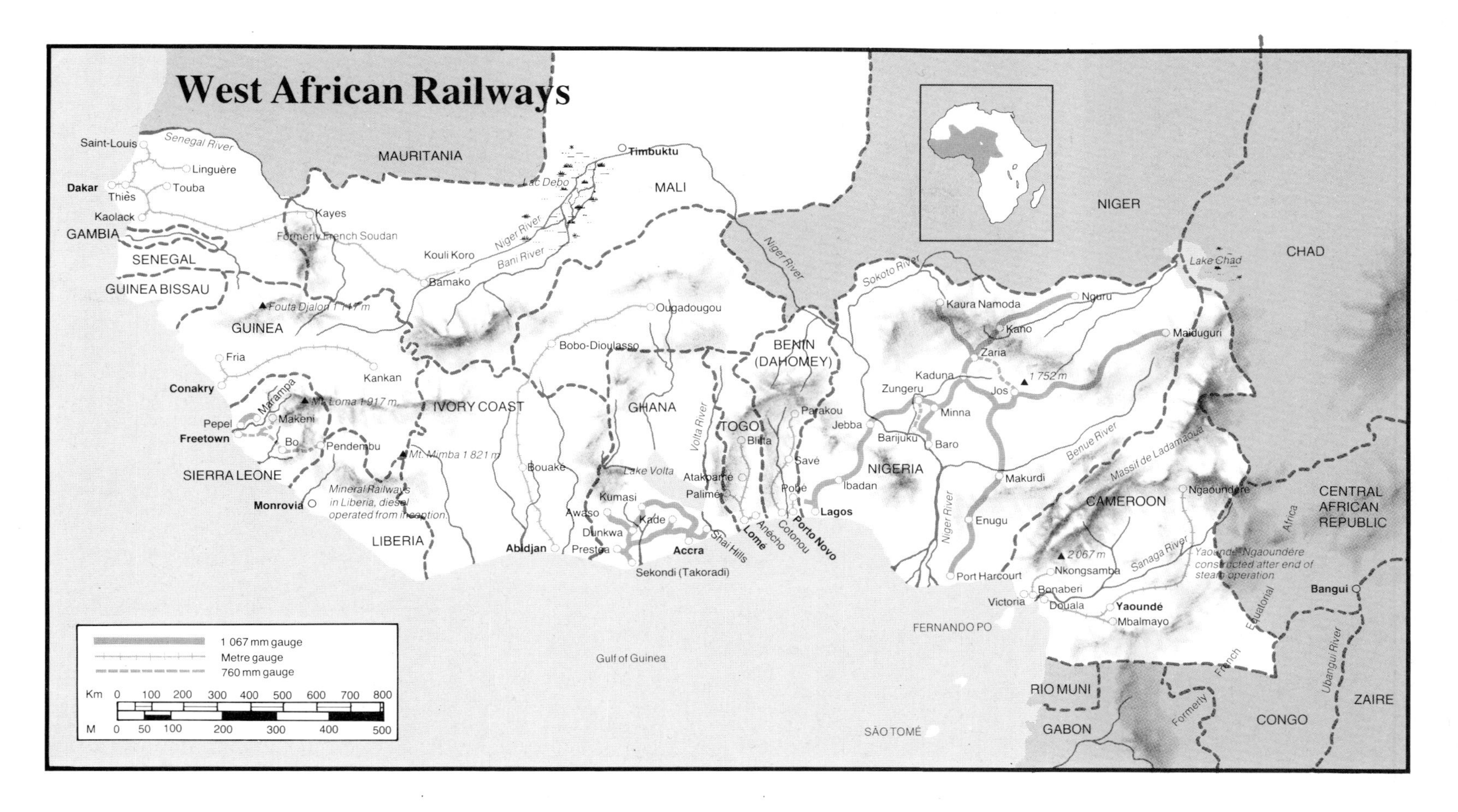

Gesellschaft which began work on the Kamerun Nordbahn in 1906. The line started from Bonaberi, on the north side of the Wouri River opposite Duala, and ran to Nkongsamba; later a branch line to Kumba was added. Of the seven locomotives built for this line, five were standard 2-8-0Ts from Orenstein & Koppel and the other two were probably smaller 0-6-0Ts used for shunting. Until 1955 when a road-rail bridge was built across the Wouri River, the Nordbahn remained separated from the main system of the rest of the country.

This was the Mittellandbahn which, despite a chequered start, was clearly intended to be more main line in character than any other German railway in West Africa. From the outset tender engines were used instead of tanks, but progress was slow – the forest saw to that – and by 1914 the line had only reached Eseka, a mere 180 km from its starting point at Duala. However, the pressures of World War I generated accelerated construction and the line was pushed through to Yaoundé, with a branch to Mbalmajo, both as 600 mm-gauge Feldbahnen.

When, after the war, this part of the territory became French Cameroun, the Feldbahn section was converted to metre gauge and extended well up-country from Yaoundé to Ngaoundéré.

The narrow-gauge locomotives were probably standard German Army types, though details are lacking. Early locomotives on the metre gauge were 0-6-0Ts, and the batch of such engines 'missing' from Tanganyika and referred to in the previous chapter was probably transferred to this line. The main-line engines were of unusual design for a colonial railway in that they were 0-8-0 tender engines without leading bogies let alone pony trucks. Presumably the existence of thousands of G7, G8 and G9 0-8-0s on the Prussian State Railway contributed to the choice. In the event, 12 were built by Borsig in 1911-12.

Following the French takeover, the railways were re-equipped with French-built locomotives – initially 11 2-8-2Ts by Corpet Louvet, of which six went to the renamed C.F. du Nord and five to the C.F. du Central. For longer distance work, particularly on the central line, the Borsig 0-8-0s were supplemented by four 2-8-2 tender engines built by Haine St. Pierre in 1930, these being the last new engines before World War II. The Mikado type then became standard, and in 1944 the line obtained ten of Baldwin's wartime MacArthur engines from the British WD. Nine further 2-8-2s from Corpet Louvet in 1949-50, were the last steam built for the country.

FRENCH SUDAN
(Senegal and Mali)

Senegal and Mali share a common rail system laid down in French colonial days, when Mali was known as the French Sudan. The first railway in French West Africa was the Dakar-St. Louis railway in Senegal. Traffic was landed at the deep-water port of Dakar, railed to St. Louis, and then taken by barge upriver to Kayes, beyond which navigation ceased. In order to serve the French Sudan, an inter-river railway was built, from Kayes on the River Senegal to Koulikoro on the Niger. This Kayes-Niger railway was opened in 1904. The ship-rail-river-rail-river route was a clumsy and expensive means of transport and a new rail link to by-pass the Senegal River was started from Thies, near Dakar, traversing Senegal to Kayes where it joined the Kayes-Niger railway. The united system, completed in 1924, was known as the Dakar-Niger railway. Over the years, some 200 steam locomotives worked on the combined systems, but dieselization commenced in 1945 and was completed by 1956.

The earliest locomotives were small 0-6-0Ts by Batignolles, built between 1883-86, and doubtless originally used elsewhere for construction work. The same firm supplied the first main line locomotives, 0-6-0 tender engines, about ten of which were built in 1893-1902. Further 0-6-0Ts were built by Weidknecht Freres in 1896-97, and another batch of 0-6-0 tender engines came from Corpet Louvet as late as 1909. All these little engines lacked power and speed, but in an undeveloped territory traffic must have been light, and even a metre-gauge 0-6-0 was much faster than any other available form of transport.

As passenger traffic developed, greater motive power was called for and

148

Weidknecht built seven 4-4-0T engines in 1901, suitable for higher speeds but rather limited in range, although probably operated with auxiliary water tanks. These were developed into a 4-6-0T, combining the 0-6-0s power with the better tracking abilities of the passenger engines, and three batches were built, each larger than the previous. First were ten Batignolles engines dated 1903, followed by ten from Decauville in 1910, both of rather curious design, with cylinders behind either a very short wheelbase bogie, or possibly a four-wheel bissel truck. Twenty-three larger 4-6-0Ts were built by Cail from 1910 to 1926, these being of more modern design and similar to the well known examples on the Reseau Breton in France.

From 1913 to 1926, 30 4-6-0 tender engines were supplied by various builders, an indication that the railway was really beginning to develop. On one stretch of the line there appears to have been a mountain section requiring assisting locomotives of greater capacity, and three 0-8-0Ts of 1910-12 were followed by four 0-6-6-0T Mallets by Batignolles, between 1910-20, again similar to those on the Reseau Breton. The completion of the Theis-Kayes link brought much extra traffic formerly sent down the river, and the 2-8-2 tender engine made its appearance, firstly with 22 Batignolles locomo-

148. A handsome 4-6-0, No. 107 of the C.F. Theis-Kayes, stands with driver and shed staff at a depot in Senegal in colonial days.

149. The last active steam in Senegal was this Belfort-built 2-6-2T, of 1893 vintage, employed by the Sodec Company at Lyndiane, near Kaolak, for shunting at their ground-nut oil factory. This ex-C.F. Departmentaux (Digoin-Etang) engine burnt ground-nut husks as fuel.

150. Across the Pont de Galougo, on Senegal's Kayes-Niger line, a diminutive 0-4-0T, probably by Decauville, treads warily with a 1903 vintage freight train.

151. 'Depart de l'express du Soudan'. A Belpaire-boilered 4-6-0 on the C.F. Theis-Niger embarks passengers and luggage for the long journey inland, all stations to the River Niger, with boat connection to the legendary city of Timbuktu.

152. Piled high with coal briquettes, a pair of 4-6-0Ts prepare to depart from Faboulima station, Guinea, with a passenger train on the Conakry-Niger Railway.

149

150

151

tives with an 8 tonne axle load, built 1923-28, followed by 16 heavier engines, with 10 tonne axle loads, by Corpet Louvet in 1926-39. During this period, Corpet Louvet also built ten 2-8-2Ts, presumably for heavier local duties, and five 0-4-0Ts, probably for dock shunting.

The largest motive power on the system comprised nine Garratts by Franco-Belge, built in 1939-40. These were more than twice as powerful as any other locomotive on the system, and had been developed from the Algerian narrow-gauge Garratts, themselves a development of the original Kenya EC class. These French colonial Garratts were of striking appearance and formed a standard design also used on the Ivory Coast. Streamlined water tanks, transverse double chimneys, and smoke deflecting plates combined to produce a unique outline. Sadly, with dieselization, these modern and powerful units lasted no longer than the older power, some of which was quite ancient when scrapped. The final locomotive classes were relatively uninspiring; 22 of the standard wartime MacArthur 2-8-2s from Baldwin were supplied in 1944-46, and finally, 15 0-6-0Ts for shunting came from Haine St. Pierre in 1947.

FRENCH GUINEA

Situated south of Senegal, much of Guinea is a mountainous plateau and the Conakry-Niger railway had to climb the ranges at the headwaters of the Niger. The railway was started from the port of Conakry in 1900 but the terrain made progress slow and it took ten years to reach Kouroussa and a further four to the terminus at Kankan. Despite its hopeful title, the line never reached the Niger, although for many years there have been projects to build a branch from Kouroussa to the Niger at Siguri, and then across the border to join the former Kayes-Niger railway at Bamako.

With fairly light traffic, the Conakry-Niger line never needed many locomotives, and in 1930 possessed only 38. Of these, the first ten, other than the inevitable construction locomotives, were 4-6-0Ts by Decauville, built in 1902-03, and numbered 1 to 10. There were some 0-6-6-0Ts transferred from the Dakar-Niger and, at a later date, some modern 0-6-0Ts by Haine St. Pierre, possibly also switched from the Dakar Line. In all probability the remaining locomotive stock was of the same basic types as on the Dakar line.

152

DAHOMÉ *(now Benin)*

Though it had two separate railways, Dahomé – as it was known under the colonial regime – is the least well-documented of the French-built lines in Africa. Both tracks were slow a-building. The Central Dahomé railway, which started out from the coast at Cotonou in 1900 took 12 years to reach Savé, some 250 km inland, though the terrain presented few problems for the builders. The line was eventually extended to Parakou, about half way up-country, in 1935.

The East Dahomé line started from the separate harbour at Porto Novo and stretched only to Pobe. Although this terminus was close to Idago on the Nigerian line, the difference in gauge ruled out any link between the two. The two Dahomé systems were joined in 1930 when the railways came under Government control and the 19 engines of the central line were pooled with the eight eastern line locomotives. These were probably all fairly standard tank engines, similar to those on other French colonial lines, but we have been unable to trace details of the exact types in use.

153. **A group of tank engines, with and without spark-arresting chimneys, congregate at the locomotive shed of the C.F. Dahomey, at Cotonou.**
154. **Battered and tatty towards the end of its working life, Ivory Coast Garratt No. 93.213, with large operating numerals '13', is attended to by shed staff.**
155. **In pristine, ex-works condition, No. 93.201 – prototype of the French West African Garratts – poses on the mixed-gauge track at Franco-Belge's works in France.**

IVORY COAST *(Côte d'Ivoire)* and UPPER VOLTA

Although the last to be started, in 1904, the railway serving these two territories eventually penetrated further inland than any other railway in French West Africa. This line was called the Abidjan-Niger railway, and, like its

156

157

western neighbour in Guinea, has yet to reach the Niger. Construction on this line was also very slow, and it took until 1913 to reach Bouaké, which, largely as a result of World War I, was the terminus until 1924. Further construction to Bobo-Dioulasso took another ten years, and the present terminus of Ougadougou was attained in 1954.

Initially there were eight locomotives, recorded as one of 17 tonnes, two of 15 tonnes and five of 35 tonnes. The two smaller types were almost certainly 0-4-0Ts or 0-6-0Ts and the larger locomotives possibly 4-6-0Ts. Decauville built several 4-6-0Ts for the line in about 1911 and these are listed in one of the firm's catalogues. The post-1924 extension from Bouaké probably resulted in some orders for typically French colonial 2-8-2s, and the climb into the highlands of Bobo-Dioulasso possibly required Mallet tank engines. In 1930, a batch of four Golwé articulated 2-6-6-2s was supplied by Haine St. Pierre for the 1-in-40 gradients encountered on this line, but later some 20 Garratt 4-8-2+2-8-4s were built for the line, to the same design and of the same batches as those on the Dakar-Niger line. In 1954, when dieselization was decided on, there were 73 steam locomotives, all of which were scrapped as the diesels arrived.

158

FRENCH EQUATORIAL AFRICA

(now Congo and Central African Republic)

For the first two decades of the present century French colonial traffic to and from Ubangui-Shari and Chad had to be carried for part of the way on a 750 mm narrow-gauge railway controlled by the Belgians. Goods and passengers could only reach the coast by boat along the Ubangui and Congo rivers to Leopoldville (now Kinshasa). From there the long stretch of river, made unnavigable by the extensive series of cataracts, was bypassed with the rail link to Matadi, whence the river again became the main artery of transport.

While Gallic pride probably played a part in the French decision to build its own line, increased trade in the territories of both colonial powers may have made a second line essential. The Belgian line had only a small capacity and it was inevitable that the harried colonial railway officials would favour traffic from their own colony when it came to allocating rolling stock.

In the event, the French proposed a 508 km railway of their own, reaching from Brazzaville, immediately opposite Leopoldville, to Pointe Noire on the Atlantic coast. It was an ambitious undertaking for, while the central section of the line followed the Niari River, between Loudima and the coast rose the rugged Mayombe Mountains which had to be crossed. Unlike the metre-gauge lines of France's other African possessions, the French Congo railway was built to 3′ 6″ (1 067 mm) gauge – probably in anticipation of a future connection with the other networks pushing upwards from southern Africa, though this was not to materialize.

The contract for the construction was awarded to the Societé des Constructions Batignolles and, though work on harbour installations began in 1921, laying of track did not begin until about 1926 and the line was completed only in 1934.

156. **For their colonial lines, the French developed a unique articulated locomotive type, the Golwé, used nowhere else in the world. This 2-6-6-2 version was built in Belgium for the Ivory Coast, where it supplemented Mallet tanks.**
157. **A refugee from World War I's trench warfare, a 600 mm-gauge former Feldbahn 0-8-0T on the Central African Republic's only railway at Mongo.**
158. **With typically Gallic flair, the only 0-4-0T on CGTA's line in the Central African Republic is numbered 020-1.**

159

160

Only one generation of steam locomotives operated prior to dieselization – a short 30 years spanning the arrival of the first construction engines in 1926 and the sale of the last steam in 1956. Surprisingly, although Batignolles built their own locomotives, they ordered their construction engines from Orenstein & Koppel who supplied six of their standard design 0-6-0Ts in 1926-27. After the line had been built these were used for shunting.

For main-line working, two types were needed, articulateds for the coastal mountain section and conventional machines for the less steeply graded inland stretch. Corpet Louvet supplied six 2-8-2 tender engines in 1930 for the easier work, and the same year saw Haine St. Pierre providing three Golwé 2-6-6-2s for the mountains. That a pioneer railway should initiate services with an untried form of articulated steam locomotive was remarkable, but the Golwé appeared to give satisfaction, for when new locomotives were needed in 1935, Batignolles supplied six shunting tanks (believed to be 2-6-0Ts and 0-8-0Ts), two more Golwés to the original design, and five more to a modified 2-6-6-4 design, with a larger tender section. One of the latter appeared on a 60 centime postage stamp issued by the Republic Populaire du Congo, a very rare type of locomotive to grace a philatelic collection.

During World War II, when this Congo-Ocean railway formed a vital link in the Allied supply lines, three of the British WD 2-8-2 + 2-8-2 heavy Garratts were allocated to the railway, becoming the most powerful engines of the system. Only eight further steam locomotives were built for the line, all 2-8-2 tender engines by SACM of Graffenstaden, France. These were of the same basic design as previously supplied to Indo-China, slightly modified to suit the wider gauge and the different coupling and braking arrangements of the Congo. Five were built in 1948, followed by three, to the same basic dimensions but modified in detail, in 1951.

The Congo-Ocean dieselized fully in 1953, making its stock of fairly modern and almost new steam locomotives redundant. The post-war 2-8-2s and the Garratts were sold to the Moçambique railways. According to *La Vie du Rail* the 2-6-6-4 Golwés were 'ceded' to Angola.

The Central African Republic is generally considered to have lacked any form of rail transport, and no mention of it ever appears in the standard international railways directories. However, an interesting article in *La Vie du Rail* describes a little railway, only 8 km long, and of 600 mm gauge, which existed from 1930 to 1962, on the Oubangui River. Equipment was apparently mainly from a Decauville line used in the construction of the Congo-Ocean railway, although descriptions of that line make no reference to such apparatus. Possibly it was supplied and used by subcontractors; in the 1920s there was a tremendous amount of World War I trench railway equipment available for this sort of work, most of which has never been traced.

As may be expected, the locomotive stock of this mini-railway, which ran from Mongo to Zinga, was as diverse a collection of second hand material as one could hope to assemble in tropical Africa. There was an Orenstein & Koppel 0-4-0T of their standard design, a Maffei 0-8-0T of German Feldbahn origin, and a Baldwin 2-6-2T. The fourth engine was also from O & K, one of their standard designs slightly modified to an 0-6-2T. The definitive history of these little locomotives is almost impossible to detail, but perhaps these scraps will act as bait for further research. This railway only ran when the river water level was too low for navigation.

159. **When the Congo-Ocean Railway in French Equatorial Africa was dieselized in the mid-1950s, several new and useful steam locomotives were made redundant. Eight 2-8-2 tender engines were sold to Moçambique, where two are shown doubleheading near Nacala in 1969.**

160. **The Congo-Ocean Railway used three of the British War Department heavy 2-8-2+2-8-2 Garratts until dieselization, when they were sold to Moçambique, joining nine similar ex-Rhodesian locomotives on heavy work on the Beira line. No. 987, representing this class, is actually an ex-RR example.**

7 THE BRITISH DOMAIN

In colonial days, where the British went railways soon followed – often into areas where transport was extremely primitive or non-existent. Certainly this was so in the three pockets of British influence along the West African coast: Sierra Leone, the Gold Coast (now Ghana) and Nigeria. Isolated from each other by the colonial possessions of other nations, their railways developed independently, though the Gold Coast and Nigeria shared a common gauge.

What might have been the first railway in West Africa was planned as a military line to provide matériel for the British forces engaged in the Ashanti War, which racked the Gold Coast during the 1870s. Material to build a railway was shipped to Takoradi, but before building began the war had ended and the immediate need for a rail link with the interior fell away.

A few years later, a grandiose scheme for a railway in Sierra Leone was proposed. Its promoters intended the Grand Sahara Railway to 'bring the company's domain in Africa to within 75 hours of London' by means of a standard-gauge railway. Though the plan came to nothing, the agricultural potential of the area was recognized and led to the construction of the first railway in the region late in the 19th century.

SIERRA LEONE

The train to Bo
She no agree to go
The engine she done tire
For lack of plenty fire
The train to Bo
She no agree to go

first reproduced in *Narrow Gauge Around the World.*
P.B. Whitehouse & Peter Allen.

Sierra Leone as well as being the smallest of the three British colonies in West Africa, had the 'narrowest' of the narrow-gauge main lines. This was the 760-mm-gauge line running from Freetown, 365 km inland to Pendembu. A 135-km branch to Makeri brought the total route distance to a round 500 km.

Sadly, this entire system, the larger of the two in the country, was closed in sections between 1971 and 1974 – its colonial past being its own greatest enemy. Built as cheaply as possible, the railway was to suffer the legacy of light construction and the resulting inability to compete with the ultimate enemy – road transport.

Even until the latter years of the last century, only footpaths pierced the jungle of this forsaken colony, which had come under British rule in 1787. At first it was used principally as a dumping ground for former slaves from England and later the Americas. Later, European colonization began – in spite of the inhospitable climate – and, as the interior had good agricultural potential, the local administrators and inhabitants pressed for a railway.

Some 20 years had passed since the Grand Sahara scheme of the mid-1870s, when, in 1895, authority finally was given to build a 760 mm-gauge railway, 50 km to Songo. This site was chosen as an interim terminus, for there was considerable controversy over the eventual direction the railway should take and its ultimate destination. From Freetown, Songo was at least in the logical direction – the rest could be settled later. Climatic problems soon manifested themselves; during the first two years of railway building, three engineers died of tropical illnesses.

When the railway was opened to the public for the first time, on May 25, 1896, the most notable event was the derailment of the engine pulling the trial train. Guests had to walk home from the celebrations – a rather inauspicious beginning.

When the question of destination had been settled, construction continued

161. **Steam in Sierra Leone lasted as long as the railway itself, but towards the end everything was in a sad state of repair. Shortly before the line's closure, a post-war Garratt, No. 64, still resplendent in slightly faded green, and a scruffy 2-6-2T prepare for another day's work at the Fisher Lane Shed near Freetown.**

162. **There can be few railways which, having been abandoned, have been resuscitated in a new location a continent away. But this was the case with Sierra Leone's railway, where an entire train – locomotives and carriages – was transported to Wales where it now runs on the Welshpool and Llanfair, having been carefully restored under the direction of Mbaja Roberts.**

161

162

163

164

beyond Songo, but was soon delayed by an uprising in the Karene district. The railway brought in supplies and ran troop trains, for by now, the line was well inland, snaking a path towards the Liberian border. Ahead now lay the Kanbui Hills – so heavily forested that it is amazing that surveyors found a route. But they did and Pendembu was reached in 1908. The final 11 km, from Baiima to Pendembu, was opened as a 'tramway', and considering the lightweight construction throughout, it is difficult to understand this differentiation, as the so-called main line, laid with 15 kg rail, was very much a tramway itself.

A branch, also termed a 'tramway', was opened to Makeni in the northern province in 1914. At first it was intended to extend this even further north to Baga, and though the line did reach Kumrabai – 35 km beyond Makeni – World War I halted construction. This was never resumed, and the extension north of Makeni was eventually closed during the early 1930s.

The final branch built was a marvel – the so-called 'mountain railway' – connecting the Water Street Station in Freetown with the newly developed European residential area high in the hills overlooking the malarial murk of the town. In 9 km the line rose nearly 250 m, reaching its terminus a mere 4 km, as the crow flies, from its start. With 1-in-22 grades, the line must have been a miniature 'Darjeeling'. Its life was short, however, and it was abandoned in 1929, a victim of road competition and providing a taste of things to come.

On a line laid with 15- and 20-kg rails and with a ruling gradient of 1-in-50 (with shorter sections of 1-in-30) and 100 m radius curves, one would expect an interesting assortment of motive power. This was indeed so, though the beginning was simple – two small, six-coupled tank engines being supplied by Hunslet in 1897. A year later, an even smaller engine, an 0-4-0 saddle-tank arrived from Bagnall. In the same year the first of a long line of 2-6-2Ts appeared. Eventually 32 of these were built by Hunslet, the first number series, 27 to 47, from 1898 to 1920, and the second number series, 81 to 85, in 1947 and 1954. These engines influenced the establishment of Bo, today the most important inland town.

In 1902 Bo was the railhead, and trains could just make the 226 km from Freetown before nightfall. When the line was extended, Bo remained an overnight stop and the town developed. This two-day operating arrangement became feature of the railway's schedule, although in later years an 'express' train traversed the entire line in one day at an average speed of 21 km/h. Considering the maximum permissible speed of 32 km/h, with numerous severe speed restrictions along the way, this was better than it seems.

Several 0-6-0Ts and five large 2-8-2Ts were bought between 1902 and 1906, and by 1910 the roster listed 32 engines. That year the railway's first tender engines arrived. These were 4-8-0s, built by Nasmyth Wilson, and by 1913 six were in service. Later batches by Hawthorne Leslie and North British brought the total to 17 by 1921. In World War II more were ordered from Barclay and Bagnall, bringing the total to 37 locomotives (nos 151-187), the most numerous single type on the system. To increase adhesive weight, one was later converted to 0-10-0, but the long wheelbase apparently played havoc with the track and was not repeated.

World War II placed a great strain on the little railway, which was still short of motive power, and the War Department supplied six 2-8-0s. These engines, built in Switzerland in 1920, but with 20 years acclimatizing in the south of India, arrived carrying the usual high WD digits of 74005 to 74010 – a vast change from their South India numbers W1 to W6. Sierra Leone eventually gave them more reasonable numbers, 111 to 116.

Final wartime arrivals were six Beyer Garratts, repeat orders of Sierra Leone's first Garratts – three 2-6-2+2-6-2s built in 1926. These engines, numbered 50 to 52, had given a good account of themselves, virtually doubling the load of smaller 2-6-2Ts and 4-8-0s, and four more were built in 1928 and 1929. After the last six were delivered in 1942-43, one engine, No. 56, was rebuilt as a 2-8-0+0-8-2 and as such could haul a load of 220 tonnes – 35 tonnes more than its sisters. Eventually four more engines were converted and their double eight-coupled configuration provided the ideal basis for Sierra Leone's final and most impressive Garratt type.

Like most other developing countries, Sierra Leone showed considerable growth after the war, and in 1949 when the railway celebrated its Golden Jubilee, plans were afoot to relay the main line with heavier rail and to realign

163. This train 'she no go to Bo' – not because of lack of steam – but because No. 45 was only shunting, at Government Wharf in Freetown back in 1940.
164. A 4-8-0, No. 178, builds up steam in preparation for hauling this long passenger train from Cline Town, a short distance from Freetown. Sunshades on the carriages are an indication of Sierra Leone's tropical climate.
165. Small, but powerful. Third order Garratt, No. 57, was one of six 2-6-2+2-6-2s built by Beyer Peacock in 1943. Altogether, 13 of these 'first series' Garratts saw service before being supplemented by the large post-war 4-8-2+2-8-4s. After the introduction of Garratts in 1926, the railway management reported, in the 1927 General Manager's statement, that by comparison with non-articulated locomotives, 'an increase of 12% in ton-mileage was obtained with a reduction of 11,9% in coal consumption per ton-mile and a 7,7% reduction in train mileage and of 3,9% in engine mileage, whilst the vehicle mileage increased by 5,7%'.

165

the Freetown-Bauya section, steepest and most difficult to operate on the system. Normally trains leaving Water Street Station in Freetown would be banked through the streets to Cline Town Station, where the railway's workshops and headquarters were located. From here the railway was on its own right-of-way, running through jungle-clad hills on a 1-in-50 grade, round a large sweeping horseshoe before arriving at Fisher Lane – later site of a goods yard and locomotive depot. Continuing on its climb along a natural shelf in the steep mountainside, with numerous views of mangrove swamps in the Sierra Leone River below, the little trains had to cross a multitude of viaducts bridging deeply cut rivers. At mile post 11, the line crossed the Orogoo River on a spidery six-span steel structure, 120 m long and 22 m high. Continuing to Bauya, more bridges, cuttings and horseshoes were encountered, and beyond, although the country was more open and less rugged, the railway continued its curving course. In the 226 km to Bo (altitude 80 m), the line climbed more than 500 m, describing 75 complete circles as it went.

With no local coal, the fleet of hungry mountain-climbing locomotives was fed coal briquettes imported from Wales. While this pleased the British exporters, it was expensive and, after World War II, the Welsh coal strikes made supplies unreliable. Local lignite was tried, but proved of little use, while Nigerian coal gave the engines digestion problems – leading the railway to think in terms of oil and, eventually, diesel.

Not only was traffic increasing, but its patterns were changing. Formerly the bulk of traffic was coastwise agricultural products, with a smaller tonnage of manufactured goods flowing inland. In tonnage, this ratio had been two to one in favour of exports, but now this was reversed. Although iron and chrome exports became significant, these were offset by up-country movement of petroleum products, additional imported manufactured items, and drop-off in exports of rice and palm oil when higher living standards raised home demand for such products. The railway was bursting its narrow gauge. Suggestions were made to convert the entire line to standard gauge, but this would have meant building a completely new railway. In the meantime orders for locomotives placed years before, but delayed by slow post-war deliveries, began to arrive and these alleviated the problem – if only for a while.

The 14 new locomotives were remarkable Garratts which, in size and power, outranged anything else seen on the railway. Weighing 70 tonnes and restricted to a 5-tonne axle-load, these 4-8-2+2-8-4s had a tractive effort of 11 000 kg (24 250 lb) and could haul a 270-tonne train up the 1-in-50 grades. By May 1956 all were in service – but also in operation were three new Hudswell Clarke diesel shunters. Small as these were, they marked the beginning of the end for steam – and the railway itself.

In 1957, eight larger main-line diesels were ordered, mainly for the rapidly-increasing passenger traffic, and in 1958 a '75% dieselization programme' for the next three years was announced. However, the latest Garratts were turning in such fine performances on goods trains it was decided to retain them, converting them to oil burning. Though the railway was turning its back on coal, steam was still part of its future plans. This proved wise, for the diesel fleet – which had grown to 33 units by 1962 with steam reduced to 30 engines – was not proving entirely reliable. Perhaps the complicated dual-fluid-drive transmissions and Paxman engines were too much for the shed and shop staff, but breakdowns, at first sporadic, were now recurrent. This gave steam an extra lease on life and older engines, due for retirement, were overhauled and returned to service.

By now other factors were emerging which would profoundly affect the future of the railway. When Sierra Leone became independent in 1961 its new administrators found democratic government a burden. By the late 1960s martial law, coups, counter-coups, and mass rioting had thrown the country into turmoil; the early 1970s saw all the trappings often associated with young developing countries – volatile politics, an unsteady economy, chaotic road transport . . . and a declining public railway.

The closure of the Makeni branch in 1968 was the start of this decline. The railway, it was argued, was inadequate for the country's needs and was blamed for many of its ills. In 1971 the main line was cut back to Kenema, and two years later to Bo. Finally, in 1974, all fell silent on the narrow gauge rails. Of 116 steam and 33 diesels operated over the years, only five steam and four diesels lasted to the end.

Meanwhile, Sierra Leone's other railway, a private undertaking, continued to operate. The Sierra Leone Development Corporation Railway, never in the limelight, was not registered as a common carrier.

Opened in 1933, it served the private port of Pepel, 25 km across the Sierra Leone River from Freetown. Built as a 1 067-mm line with 32-kg rails, allowing a 13-tonne axle-load, this railway ran to large iron ore deposits at Marampa, 80 km inland. The line was powered by four Garratts, all 2-8-2+2-8-2s – in design virtual duplicates of South African class GE Garratts – and they worked 1 300-tonne loaded ore trains on 1-in-125 grades. Apart from some shunting tank engines, they monopolized all the work and the line was described in Durrant's *The Garratt Locomotive* as being 'the only railway in the world where all traffic between inception, to the end of steam was hauled by Garratts'. In the late 1950s diesels took over and, though the steamers were held in reserve for a time, they were eventually scrapped.

Even this railway may have disappeared, for the SLDC went into liquidation in 1975, and the railway has been used infrequently, according to latest reports.

The demise of railways in this West African country is tragic – all the more so since so few people photographed it – but there is happy consolation in the fact that one engine, 2-6-2T No. 85, last of the 1954 Hunslets, together with four coaches has been preserved – in Wales, of all places. The Welshpool & Llanfair Light Railway Preservation Co. purchased this complete train, which arrived at Liverpool on August 7, 1975, and today can be seen in operation many miles from its original habitat. For No. 85, the country she runs through is just as geen – although a bit on the cold side – but at least she has come home.

166 167

168

NIGERIA

'The train carrying the Prince of Wales on his tour of Nigeria arrived at Minna at midnight on April 17, 1925. In the afternoon the Prince drove the train for half an hour. The action of the Prince prompted senior railway officers to present him with the mileage sheet, showing that the total amount due to the Prince was 10d. The Prince smilingly signed the sheet and accepted the money.' from *The Railway Gazette*.

In 1925 such an event was newsworthy, and Nigeria was a British colony. Nigeria gained its independence in 1960, and is rapidly moving ahead – the 'big brother' of the three former colonies. It covers a large area and its 68,5 million population, is the largest in Africa. And while its 3 505 route-kilometres make it the largest rail system in West Africa, on the grand map of Africa this is meagre.

The British landed at Lagos in 1851, intending to stop the slave trade, and ten years later the area was ceded to Great Britain, becoming a colony. Rivers were the main routes for the produce of the interior but Lagos, with its good natural harbour, was located far from the mouth of the Niger – the main artery of trade. Clearly, if Lagos was to prosper, good communications with the interior were essential and the late 19th century answer for such development was to build a railway. A start was made in 1896, from Ebute Metta on the mainland opposite Lagos Island, and by March 1901 the 1 067 mm-gauge line was opened 200 kms inland to Ibadan.

Simultaneously, other railway development was taking place deep in the interior at the established centre of Zungeru. This town was chosen as a seat of the expanding colonial government, and to connect with Barijuku – head of navigation on the Kaduna River, a tributary of the Niger – a 760-mm 'steam tramway' was opened in 1901.

Construction on the Lagos line resumed in 1904, and by 1909 had reached Jebba, 500 km inland on the southern banks of the Niger. Here the lack of a bridge delayed progress but already, by penetrating deep into the interior and reaching the Niger, the railway provided Lagos with much trade.

Then Sir Percy Girouard became High Commissioner for Northern Nigeria. He had been the director of the Sudan Railway and President of the Egyptian Railway board. In South Africa he had been the British army's railway director during the Anglo-Boer War, afterwards becoming Railway Commissioner. Clearly a man for railways, soon after his arrival in 1906 he supervised a new line. This began well inland, at Baro, 400 km up the Niger, and 200 km below Jebba. Its goal was the large northern town of Kano – the walled city of the Hausa and starting point for Saharan caravans. Construction was pursued vigorously, and this 570-km line, built to a gauge of 1 067 mm, was completed by 1911.

At Jebba, the line for Lagos was being extended across the Niger, first by a ferry, then on two bridges, the second of which was opened in 1916. With the river crossed, controversy arose as to which gauge to use on the extension. Originally 760 mm had been proposed, to correspond to the Zungeru line's gauge, but as it had been decided to continue on to Minna on the Baro-Kano line, 1 067 mm was chosen. In 1912 the new line reached Minna, and after the South Channel Niger Bridge was built, through-traffic could flow uninterruptedly from Kano to Lagos. This railway achievement, as much as anything else, helped unify Nigeria.

In the interior, north of the tropical forest zone, development of agriculture and then mineral resources was rapid. Tin mining began on the Bauchi plateau, to the south-east of Zaria, and a 760 mm-gauge line was started in 1911, reaching Bukuru in 1914. Material for this railway came from the old Wushishi Tramway at Zungero, which was uplifted in 1910.

The discovery of coal was the next spur to railway construction – this time in the south-east at Enugu. A good potential harbour had been found 60 km up the Bonny River to the east of the Niger River mouth, and a 320-km railway was built linking the harbour, later named Port Harcourt, and the coalfield. This was opened in 1916, allowing the railway to use local coal for the first time, but as coal still had to go by sea to Lagos, a rail connection with the rest of the system was vital. This was only completed in 1927, and even then a ferry across the Benue River at Makurdi had to serve for five more years.

Some short branches were built in the 1920s, but further rail development came to a standstill for more than 25 years. Then, in 1956, a decision was made to construct a 670-km line from Bukuru to Maiduguri in the north-east. This line, opened in 1964, was the last built until recently.

Nigerian locomotive design in the mid-1920s was quite uninspiring. Perhaps the tropical climate – hot and humid enough to debilitate all but the most determined engineer – had an effect, for the various early locomotives were nondescript and ineffective. As in Ghana, the first locomotives were light construction types and these were followed by odd, non-standard engines, including a 2-8-0 design copied from Ghana.

Problems not confined to Nigeria, were poor footplate staff performance, lacklustre maintenance, and the high proportion of engines awaiting repairs. In 1924, a report entitled *The Railway System of Nigeria,* by Lt.Col. F.D. Hammond of the Royal Engineers, took a dim view of the situation and suggested ways to improve efficiency. The personnel problem was most acute as excerpts from his report suggest:

'At the beginning of this year the establishment of main-line drivers was made up as follows – 65 Europeans, 14 West Indians, and 12 Natives. In addition there are 25 Native shunting drivers . . . This policy of relying practically entirely on imported drivers has been adopted by the Nigerian Railways ever since the early construction days, when of course their employment was essential. An attempt was made at one time to train 12 Native apprentices, but these apprentices did not apparently take kindly to the work of firing which they regarded as a labourer's work and the scheme was dropped. Apart from this short lived experiment, no systematic attempt had been made and the term systematic is hardly applicable to this isolated case. It is in marked contrast to the procedure in all neighbouring English and French countries where the European driver is now the exception and not the rule.'

Of the engines themselves, Lt.Col. Hammond noted that '157 main line locomotives are on the books, consisting of 13 classes.

'One class in particular, the 101 class 2-8-0s (Gold Coast type), total 25 on the books; and none are employed on the main line and only three in shunting work, the remainder being laid up. In the class, the leading pony truck and engine axle boxes are badly designed; the wheels are small and unbalanced. If run at any rate of speed, they knock themselves about and have to come back into the shops for heavy repairs at frequent intervals, besides being a constant source of trouble in the running sheds.'

Apparently the only really effective engines were the 255 class engines, consisting of several sub-classes of 4-8-0s dating from 1909 and including some Nasmyth Wilson engines constructed in 1923. In most of Africa, where 1 067-mm and metre-gauge railways were constructed quickly and cheaply, generally only engines with four-wheel leading trucks were successful. The 4-8-0 became the most successful narrow-gauge locomotive, until the introduction of the 4-8-2. A close look at the specifications of Nigerian, Ghanaian and Nyasaland locomotives, shows that their designs were based on the formidable Cape Government Railways' 7th Class, first introduced in 1892. This sure-footed 12-wheeler had small 1 086 mm (3′ 6¾″) diameter coupled wheels, while a light axle load of 9 tonnes permitted operation on rail as light as 22,5 kg (45 lb). Unsuperheated steam was fed to slide valve cylinders, measuring 432 mm x 584 mm (17″ x 23″), and inside Stephenson valve gear was employed. Long after the countries of the southern sub-continent had discarded this 'colonial' type, refined examples with superheating, piston valves and Walschaerts valve gear were still built in West Africa. When larger engines were introduced, Mountain type 4-8-2s, remarkably similar to their southern counterparts, appeared.

In Nigeria, the jump from small to much larger engines came quickly and some unorthodox designs appeared. The largest 4-8-0s, constructed in 1923, had a mass of 86 tonnes with tender and had a tractive effort of 23 600 lb (10 700 kg). These were superseded by 29 Class Armstrong Whitworth 4-8-2s, first built in 1924, which had a mass of 111 tonnes and had a tractive effort of 29 470 lb (13 360 kg). Two years later, Vulcan delivered five massive Class 801 three-cylinder 2-8-2s, with a mass of 127,5 tonnes and having a tractive effort of 38 570 lb (17 490 kg). These were followed, in 1930, by three very large 806 Class 4-8-2s, also three-cylinder machines, which had a mass of nearly 141 tonnes, and exerted a tractive effort of 39 860 lb (18 070 kg). Vulcan-built, these engines were the heaviest and most powerful non-articulated engines to run in Nigeria. These two classes of Vulcan-built engines were restricted to 40 kg rail and, although successful, they were not repeated – their service being limited to the Lagos-Jebba main line.

An experiment, unique to the 1 067 mm gauge in Nigeria, was the fitting of boosters to two of the 1924 Class 701 4-8-2s, which entered service in February 1929. With these trailing truck boosters, a load of 570 tonnes could be hauled on the Mada-Kafanchan section of the Eastern Line from Port Harcourt, compared to 460 tonnes for a standard 701 Class engine. However, such mechanical contrivances needed extra maintenance; when seen in later years not only were the boosters removed, but the three-cylinder engines had been converted to two-cylinder by removing the inside driving rod and the Gresley conjugated gear.

For 'express' passenger services in the 1920s, ten Pacifics were built by Nasmyth Wilson between 1926 and 1928. These fine Class 405 engines had narrow fireboxes, placed between plate frames, and a lighter axle-load than the large Mountains and Mikados, enabling them to operate on 30-kg rails. Though widely-used, their 1 524 mm (5′) coupled wheels could not have been fully exploited as the maximum speed – even on 40-kg rail – was restricted to

169

170

171

166. **The Baro-Kano railway, opened in 1906, was completely isolated until 1912 when it was linked with the Lagos railway. Previously it had operated its own locomotives including this R. Hawthorn-Leslie 2-8-0, named 'Bacdecci'.**
167. **This large Belpaire boiler 4-8-0 of Class 301 was built by North British.**
168. **Nasmyth-Wilson & Stephenson built nine of these large 4-6-4Ts between 1927 and 1930.**
169. **Small but relatively modern, this Bagnall 0-6-0T, No. 18, is one of 30 of its type built between 1923 and 1928 and numbered 1-30.**
170. **From its opening until 1907 the Lagos Railway operated 36 steam locomotives of seven classes including No. 21 which was one of six 2-6-0s built by R. Hawthorne-Leslie for the opening of the line in 1898.**
171. **No. 203 of the Lagos Government Railway was a 2-6-2, built by Nasmyth-Wilson. The unusual Nigerian feature – also to be found in Ghana and East Africa – of a curved cab-side was needed for running clearances.**

172

173

70 km/h. But then, even on the principal passenger trains such as the 'North Mail' or 'Boat Express' from Lagos to Kano and the 'Ocean Mail' in the return direction, the average speed was only about 34 km/h.

The problem of light rail convinced the Nigerian railway authorities that the solution lay in articulated locomotives. As they were thinking 'big' in the late 1920s, two large 4-8-2+2-8-4 Garratts were ordered in 1930. These were intended for use between Jebba and Minna, and from Zaria to Kano on the Lagos-Kano main line. Large, non-articulated engines would work through loads over the 40-kg rail to Jebba and Garratts would work the 22,5-kg rail sections. In practice, it was found that the Garratts were too powerful and the designers went back to their drawing boards. A three-cylinder 2-10-4 was considered, but Beyer Peacock convinced the railways that another Garratt design would be best, and in 1935 the first of a group of 22 Class 501 4-6-2+2-6-4s was delivered.

Many colonial railways ordered locomotives, took them for better or worse – and used them until they fell apart. Often, design faults were disregarded and engines had to soldier on as best they could. This was not so in Nigeria, where an extensive rebuilding programme began in the 1920s and continued into the 1940s. Ten of the much maligned early 2-8-0s were converted to 0-8-0Ts, while several classes of 4-8-0s were rebuilt with superheated boilers and Walschaert valve gear. A more dramatic change was in store for the 401 Class express passenger 4-6-0s which were rebuilt as handsome 4-6-4Ts. Finally, in the 1940s two of a group of 27 1925-built 4-8-0s, which had suffered chronic steaming problems, were reconstructed with modified steam pipes, cylinders and blast pipes – making them much more effective steamers.

After World War II, one traffic feature dominated the minds of Nigerian railways personnel as well as British politicians and newspaper reporters. This was the movement of groundnuts from the north of the country to the port at Lagos. As a result of wartime requirements deferring maintenance, out-of-order locomotives rose to 35% in the early post-war years. As the groundnut harvest increased in size, the available fleet diminished and this became a political issue.

To help the expected post-war surge in traffic, an order had been placed with Vulcan in 1944 for 20 medium-weight Mikados, later to be known as the 'River' class. British loco deliveries were notoriously slow at the end of the war and, in desperation, 24 2-8-2s were ordered from Montreal. Deliveries of these Canadian engines were rapid, and the first arrived in 1947. Known as the 'Newfoundland' class, these engines were similar to those running on the Canadian national railways' 1 067 mm system in Newfoundland, and were well liked by everyone. A batch of 0-8-0T+T engines, built by Hunslet and similar to those built by Hawthorn-Leslie in 1925 for Nigeria (and later by Hunslet for neighbouring Ghana) was also popular.

Eventually, in 1948, the first of the Vulcan 2-8-2s arrived and immediately were pressed into moving the groundnut harvest. Annual reports of the time indicate the acute scarcity of motive power that had developed. Altogether 79 River class 2-8-2s were ordered, the last arriving in 1956. Earlier in the 1950s a large-scale reboilering programme was undertaken, involving 69

174

175

176

172. **Three-cylinder 4-8-2, No. 807 'Sir Graeme Thompson', as built, and seen leaving Ebute Metta junction on an up-goods.**
173. **The original three-cylinder 4-8-2, No. 806, sits at Ebute Metta shed in 1964, looking very different from its original state – with smoke deflectors and the third cylinder inoperative, the Gresley conjugated valve gear having been removed.**
174. **The largest steam locomotives to run in Nigeria were two 4-8-2+2-8-4 Garratts, built by Beyer Peacock in 1930. With a tractive effort of 39 920 lb (18 100 kg), even on a mere 9,5-tonne axle-load they were the most powerful engines ever to run here.**
175. **Photographed at the Hunslet Works at Leeds, this 0-8-0T+T was one of 30 such locomotives built between 1925 and 1949 and officially classified 'mixed-traffic' machines. The basic design appeared also in Ghana, Zambia and South Africa.**
176. **When British builders were unable to cope with orders immediately after World War II, the Canadians stepped in, supplying 24 'Newfoundland' Class 2-8-2s. No. 755, 'Vancouver', was photographed at Ebute Metta in 1964.**
177. **Ultimate Nigerian steam locomotives, the River Class 2-8-2s were considerably lighter than the earlier 2-8-2s and 4-8-2s. They were, nevertheless, a useful utility locomotive. River 'Chanchaga' is a Henschel product.**

177

178

179

180

181

engines of the 4-8-0, 4-8-2, 4-6-2, and 4-6-4T types. Nevertheless, the out-of-service ratio remained unacceptably high, especially when widespread mechanical problems with the River class 2-8-2s resulted in the simultaneous removal from service of 20 with boiler stay fracture and piston valve problems.

This state of affairs played directly into the hands of the diesel salesmen who, when travelling to Ghana, visited Nigeria. This paid off, and the diesel people went home with an order for ten 750 hp units, identical to those ordered for Ghana. Delivery was to begin in 1954.

Road transport was eroding the railway's position, and on September 30, 1957, the colourful Bauchi Light Railway was closed. As its end neared, its locomotives consisted of two classes – three Hunslet 0-6-0Ts and seven Kitson 0-6-2 tender engines; most were cut up by 1960, but one of each type has been preserved at Jos. Apart from the early Wushishi Tramway, Nigeria had only two other 760-mm lines, the Lagos Steam Tramway, and of all things – the Lagos Sanitary Tramway. This railway had one locomotive known as the 'New Sanitary', leading one to wonder what the earlier locomotive, officially named the 'Kokomaiko', might have been called by the staff. Perhaps, the 'Old Sanitary', or even the 'Soiled Sanitary' . . . This could throw new light on the origin of the gricing term 'Bucket Bahn'.

From a peak of 274 steam locomotives in 1955-56, steam diminished as the diesel fleet grew. Then in 1967, seven years after independence, civil disorder overtook railway events. The country was fragmented when the eastern region seceded, forming the state of Biafra. Eighty steam and two diesel locomotives were commandeered by the Biafrans and when, in 1970, the civil war ended, much of the equipment and permanent way had been badly damaged. However, reconstruction was swift and it was not long before a total 'steam-elimination' programme had been announced. Nigeria, by now a major producer and exporter of oil, could well do without its 'old-fashioned' steam locomotives burning its inferior grade coal.

By 1975 there were less than 100 active steam locomotives and these have been further reduced. However, recent deliveries of diesels intended to bring steam to an end, have not done so. The chronic problem of large numbers of locomotives being out of service has also affected the diesel fleet; whereas most steam spares could be manufactured locally, diesel spares must come from overseas – so many diesels remain idle. Recent plans call for an end to steam by 1981 or 1982, and only time will tell if this deadline can be met.

Recently Nigeria's 1 067 mm-gauge system was declared unable to cope with the country's future requirements and a decision was taken to convert to standard gauge. The first section of this new gauge – which will be from Port Harcourt to Ajaokuta, the site of a new steelmill – is expected to be completed by 1983.

There is a flurry of activity and groups of foreign 'advisors' have assessed the needs of these new railways. It is good to see railways figuring in Nigeria's future, but forgotten are the locomotives which 'pioneered' this large country, the locomotives – some only 23 years old – which have been declared 'obsolete'.

178. **When clean – a rather graceful engine. No. 409 'Oba of Benin' is just outshopped at Ebute Metta on November 2, 1964.**

179. **'The train that stands on the platform' is how R. Geunin, the Nigerian Railway photographer, described his photograph of Bauchi Light Railways' 0-6-2 No. 60, ready to leave Jos for Zaria soon before the line closed on June 20, 1952.**

180. **On parade at Ebute Metta shed are (from left to right) No. 42, a 2-6-2T, No. 771, a Montreal 2-8-2, No. 405, a Nasmyth-Wilson 4-6-2, and late model 'River' Class 2-8-2 No. 114.**

181. **Christmas in Lagos. This 0-6-0T, decorated for the event, carries a jovial Santa Claus on his annual journey through the heat of the tropics.**

THE GOLD COAST *(now Ghana)*

In contrast with Nigeria, the Gold Coast, now known as Ghana, has a railway system restricted to the southern third of the country – roughly the area of greatest development. Geography and topography are obvious reasons for this, as the Volta River, served by tributaries in Upper Volta and the Ivory Coast, cuts a wide path right through the country – dividing it neatly into three.

Aptly named the Gold Coast until independence in 1957, its magical and alluring name was associated with the riches, mainly gold, which were found in the interior and guarded by a 250-km-deep belt of dank, thick jungle. Some of this gold was traded and in 1471 the Portuguese had already explored the area and established coastal forts. Later, by the beginning of the 18th century, several other European countries had also established forts and attempted to trade. To their frustration, they soon discovered that most of the gold was traded to the north across the deserts to the Arab countries.

However, another lucrative trade developed at the coast – the export of slaves. This continued until 1821 when Britain took control and banned the trade. The next development was a backlash by the Ashanti who were very involved in this trade in human cargo. In the 1870s they clashed with the British, who were expanding their colonial jurisdiction, and it was this 'war' which nearly brought the Gold Coast its first railway.

Sir Garnet Wolseley, later to leave his name in South African history, was Governor and Commander-in-Chief. Aware of the need for improved supply of matériel to the interior, he proposed a military railway. Supplies were brought in, but before the project began the conflict ended and the railway was stillborn.

By the late 19th century, prospectors ranging the jungles in search of gold, found it at Tarkwa, 70 km inland from Takoradi, on the coast otherwise known as the 'white man's grave'. Development of this promising goldfield required adequate transport and to this end a railway was proposed and the idea quickly accepted. Construction began in August 1898, only to be delayed by yet another war with the Ashanti. Progress was slowed, and the Gold Coast's first railway was opened in 1901.

Following typical colonial practice, the 'mother country' supplied equipment, and the first two locomotives were Fowler 0-4-0Ts for construction work. They were followed a year later, in 1899, by four larger 0-6-0Ts from Hunslet. These small engines were suited to construction, but to operate the line larger locomotives were required. R. & W. Hawthorn Leslie produced these in 1901, all tender types, comprising four 2-6-0s and five 2-8-0s. These were followed by two larger 2-6-0s in 1903.

For political and economic reasons the line was extended to Kumasi, capital of the Ashanti region, and opened in October 1903.

The jungle proved to be the greatest obstacle. Surveyors found that freshly-cut pegs, used for locating the right-of-way, soon sprouted leaves, becoming indistinguishable from the surrounding forest. This problem was overcome by staking with concrete pegs. Another problem soon to show itself was the ravages of white ants, which developed a taste for the wooden sleepers. The solution here was to use steel sleepers, and about the only organic material allowed on the railway – apart from human beings – was the wood fuel for locomotives, no known substitute being readily available. With the opening of the Kumasi line, the Gold Coast had 275 km of 1 067 mm-gauge railway, on which 19 steam locomotives were operating. More than 16 000 men had been needed for construction work, carried out almost entirely with manual labour.

The next objective was to connect Accra, capital of the colony, with the rest of the system and, though proposals to build such a line had been made as early as 1905, construction began only in 1911 and was not completed until 1923. In the meantime, one of the few 'true' narrow gauge railways in the Gold Coast was completed. This was the 760 mm Accra-Weija Tramway, opened in 1916 to serve the waterworks 16 km from Accra. This quaint little line, operating three 12-tonne Avonside 0-4-0Ts, closed in 1941.

With the Accra-Kumasi line completed the Gold Coast system was consolidated, though further development lay ahead. As in neighbouring Nigeria, engines with four-wheel leading trucks were suitable for the curvacious jungle railway. The arrival of the first nine 4-8-0s in 1909 was the beginning of effective motive power on the young railway. Between then and 1921 a further 17 4-8-0s were bought from British builders, but a few odd types also entered service, including four 2-6-2Ts and four 0-6-0Ts used mainly for shunting.

In 1912 North British delivered two relatively large 4-6-4Ts, fine-looking locomotives which, unfortunately, were not ordered again. Twelve years were to pass before more tank locomotives were ordered, these being four 0-6-0-Ts for construction on the Central Province Line. This railway eventually provided a more direct route between Takoradi and Accra, and construction began in the west from Huni Valley, extending 160 km to the east at Kade. Work stopped in 1927, when the line was completed, just 50 km short of a connection on the Accra-Kumasi line. However, would-be through passengers had to wait until 1956 to see the gap closed.

As in Nigeria, the 1920s saw rapid development in the Gold Coast, and larger locomotives, which would set the standard to the end of steam, arrived in the country. To develop a new deep-water port at Takoradi, a new 10 km-long railway was built and to help move heavy loads of granite from a quarry near Tarkwa, six large 2-8-2s were ordered from Nasmyth Wilson in 1923. These engines were so successful that eight more were ordered the following year.

1924 saw the arrival of five 'express passenger' 4-6-0s which were later named after governors of the colony. The first four of the Mountain type 4-8-2s by Vulcan also arrived in this year, to move manganese traffic originating at Insuta. These large engines were limited to 40 kg rail, and during the 1930s, when heavy rail was laid north to Dunkwa, six more were bought. Unlike Nigeria, where even larger non-articulated engines were developed, these were the largest 'straight' engines used in Ghana and were virtually identical to the Armstrong Whitworth 701 class 4-8-2s in Nigeria, except for fractionally larger cylinders and a higher tractive effort.

Another design shared by the Gold Coast and Nigeria was a series of 0-8-0T shunting locomotives, 13 of which were built for the Gold Coast between 1930 and 1936. More road-designs appeared during the 1930s, the most impressive of which was a class of seven large-wheeled 4-6-2s. These Pacifics were built by Beyer Peacock in 1938-39, having 1 524 mm (5′ 0″)

183

coupled wheels, and had similar dimensions to the ten engines built for Nigeria in the late 1920s, but were 'modernized' with bar frames and a larger and wider over-the-frames firebox. These attractive engines supplemented the earlier 4-6-0s, and were beautifully painted: 'brick red' with the Gold Coast coat of arms on the cabsides. Until the advent of diesels they were the top-link engines in Ghana pulling 315 tonne (13 coach) trains on the Sekondi (Takoradi) Kumasi line, and 300 tonne (12 coach) trains on the Accra-Kumasi line. Although relegated to local services, some were in operation as late as 1977.

For service on the Accra-Kumasi line, a series of lighter 4-8-2s known as the 'Four-Foot' 4-8-2s (for their coupled-wheel size) were built between 1936 and 1939 – the forerunners of the final steam design for the Gold Coast. During World War II, the American Locomotive Company (Alco) supplied 11 of their standard WD 2-8-2s which, though light and exerting a tractive effort of only 9 117 kg, were robust and popular. Also liked were the whistles – medium deep-toned chimes – rich in note when compared to the shrill, high-pitched, typically British whistles on all other engines.

184

The only Garratts to run in Ghana were, like the American WD 2-8-2s, delivered during the war. Purely freight locomotives, these six standard Beyer Peacock WD 2-8-2+2-8-2s hauled 30 loaded bogie wagons on the manganese trains from Insuta to Takoradi – a gross tonnage of 1 100 tonnes. Their load could have been increased but for the limitations of the antiquated central buffer and screw couplings. Unfortunately their lives were not long, being cut short by dieselization; they were withdrawn from service in 1960.

After the war, an improved version of the original 4-8-2 type (the 221 Class) was proposed, but even minor design modifications would have delayed deliveries by as much as two years, so the original design was duplicated and 15 of these engines were built in 1949. In sharp contrast to this 25-year-old design were the 30 engines of the 248 Class, built in 1951. These 4-8-2s were the last new steam engines to arrive in the Gold Coast – being an improved version of the successful 'Four-Foot' and equipped with Timken roller bearings on all wheels and rods, including the tender. In this respect, they were the only reasonably modern steam engines to operate in the Gold Coast, running up to 400 000 km between general repairs. A comparison with the most modern Nigerian locomotives is interesting: these were the River class 2-8-2s which lacked roller bearings, and though they were slightly heavier and had a slight edge on tractive effort, their marginally lower adhesion, coupled with plain bearings, reduced their effectiveness – particularly when starting heavy trains.

In 1954, the last year of full steam operation, 141 locomotives were on the

182. **A WD 2-8-2+2-8-2 Garratt hauls a heavy train of manganese ore near Takoradi.**
183. **Its original brick-red colour scheme long since a memory, this Pacific, No. 366 was photographed at Dunkwa in July 1976. Downgraded from her earlier main-line duties, No. 366 found employment on the Awaso branch line.**
184. **Another Ghanaian version of a Nigerian locomotive type, the 0-8-0T+T No. 37, is a straight tank locomotive, and was the Accra station shunter on March 4, 1973.**

185

188 189 190

186

187

191

books, compared with 83 in 1935. Of these, 62 were 4-8-2s, making the Mountain type the most popular wheel arrangement. Problems with coal supplies became acute in 1951 with a strike at the Udi mine in Nigeria – main source of the Gold Coast's coal. As a result the administration took a serious look at alternatives and, typically, diesels were seen as a viable answer. Ten were ordered for delivery in 1954. At the same time, 34 steam engines were converted to oil burning and, though this was considered a reasonable success, no more were converted. Instead, more diesels were ordered, and by 1968 the railway could report that 'all express and main-line passenger services were diesel'. In the same year one of the 248 Class 4-8-2s, No. 260, was equipped with a Giesl ejector. Since then several others have been similarly equipped, however this has not delayed the switch from steam to diesel.

Several announced dates for full dieselization have been made and none kept. The most recent published figures (for 1975) show 88 diesels, 61 coal-burning steam and 29 oil-burning steam locomotives, although only a fraction of the steamers are reported to be in regular service. Nevertheless, it is apparent that a developing country was given a good 'sales-line' with extended credit, and told how few diesels would be needed to replace the 'old' steam fleet. Good on paper, but in practice, once diesels come in for repairs and foreign exchange is tight, they sit – and steam comes to the rescue.

193

194

185 (Previous page 185-191). **Mainstay for many years of the Gold Coast Railway was the 4-8-0, 24 of which were built between 1909 and 1921. No. 172 is one of the last, and was constructed by R. Hawthorn-Leslie in 1921.**

186. **In recent years several of the post-war 249 Class 4-8-2s were rebuilt with Giesl ejectors – more efficient than the old blast pipe arrangements – but this did not stop Ghanaian firemen from making excessive black smoke.**

187. **Brick-red was once the hallmark of well-kept Ghanaian engines, but the former spruceness has faded like the paint of this 4-8-2, No. 260, leaving Accra.**

188. **A crossing at Papase between an incoming train with doubleheaded 4-8-2s and a single 4-8-2. Ghanaian locomotives operated without cowcatchers – making for a hazardous run, as much of the right-of-way was unfenced.**

189. **A post-war 4-8-2 of the older 221 Class waits at Accra with the Kumasi passenger train in December 1967.**

190. **Mr E.K. Badu, the running-shed foreman at Achiasi Junction, is particularly proud of No. 187, 'Sir Shenton Thomas', one of six good-looking 4-6-0 express passenger engines built by North British in 1932.**

191. **The chunky lines of No. 138, one of the series of heavy Vulcan-built 4-8-2s which were the largest non-articulated locomotives in Ghana.**

192. **This etching first appeared in the *Locomotive Magazine*, to illustrate the R. Hawthorn-Leslie 2-8-0 No. 103.**

193. **Hunslet built this small-wheeled 4-6-4T in 1914, similar to a number of North British engines constructed in 1912.**

194. **Steam locomotive sheds can be neat, tidy and even clean, but this Ghanaian shed is the other extreme.**

CF.A 61

SOUTHERN AFRICA

8 ARTICULATEDS IN ANGOLA

Eucalyptus-burning Garratts up the escarpment

You may have seen a Duchess on Shap or a Chapelon Pacific on Caffier Bank, and may even have seen a Big Boy on Sherman Hill; but to appreciate a steam engine in all its fiery glory, you must see a wood-burning Garratt at night, shooting an orange-red comet's tail of glowing char from its chimney as it pounds towards you like some latter-day flame-throwing dragon.

Such sights were common on the Benguela Railway, which operated one of the largest fleets of wood-burners in the world in the Portuguese colony of Angola. Since independence in 1975, parts of the country have been affected by guerilla activity; railway lines have been damaged and services interrupted, and one can write with certainty only about the position before that date.

Though a vast territory about the size of France, Spain and Great Britain combined, Angola is served only by about one kilometre of railway for every 500 km² of land. Until 1975, when the 600 mm-gauge line from Porto Amboim to Gabela was closed, there were four rail systems in Angola, none physically connected to each other. This isolation gave Angolan railways a special allure for, even though three were state-owned, there was little duplication of equipment.

In 1963 the state-owned Luanda, Amboim and Moçamedes lines were amalgamated to form the Caminho de Ferro de Angola (CFA), but the Caminho de Ferro de Benguela (CFB) continued in private operation – and has continued to do so, even after independence.

Caminho de Ferro de Benguela, or Benguela Railway (CFB)

The Benguela, with 1 347 km of track, is by far the longest railway in Angola; it is the only one to boast a connection with the rest of southern Africa; and it was also the only one privately owned. Until 1972, it was almost entirely steam operated, yet it posted an annual operating ratio in the low 60s, a cheerful figure for the shareholders of this Lisbon-based company whose shares are quoted on the London Stock Exchange. At that time the CFB had a stock of four small diesel dock shunters and 108 steam locomotives, of which nearly 80% were wood-burning. Feeding them kept a lot of men busy: 3 000 men in the CFB's own forests felled 16 million Australian bluegums every year, more than two million of which found their way into locomotive fire-boxes.

But wood was not the CFB's only fuel; more than 20% of its locomotives burned oil or coal, and in its early years even local peat was used until it was found to be uneconomical.

On the flat coastal plain between Lobito and Benguela, coal burners were used – local passenger and freight trains were hauled by Baldwin 4-8-0s of class 9B, while through-trains were generally handled by class 11s, the only CFB 4-8-2s. Oil-burning Garratts took over between Benguela and Coruteva, the demanding first stage of the climb up the escarpment. Between Coruteva and Cubal both oil-burning and wood-burning Garratts were used – at Coruteva the oil-fired helper usually gave way to a wood-fired one, while the oil-fired lead engine carried on to Cubal, a division point and engine terminal, where all engines were changed.

East of Cubal only wood-fired power was used, which usually meant a wood-fired Garratt up to Nova Lisboa, at an altitude of 1 800 m, another division point and location of the main workshops. From here the line undulates gently, gradually losing altitude until reaching the Zaire border at Texeira de Sousa. Along here traffic was mainly in the hands of North British 4-8-0s of class 9A and 9C, and of classes 10 and 10E Garratts.

It was early realized that Lobito was an ideal natural harbour, and as early as 1895 the Portuguese government granted a concession to a private company to build a railway from Lobito, via Benguela, to the Rhodesian border. After completing only the 23 km to Catumbela, the company ran into financial difficulties and its metre-gauge line ceased operations. Then, in 1899, a concession for the building of the line from Lobito to the Congo border was granted and a survey made for a route through the important trading centre of Caconda.

In the same year Robert Williams's Tanganyika Concessions was given the mineral rights in Katanga by King Leopold, after the eminent Belgian geologist, Professor Cornet, had visited the area and declared the copper deposits there to be uneconomic. The true extent of the deposits was discovered in 1902 and Williams saw Lobito as his company's natural outlet to the sea, just as he had seen Beira as the British South Africa Company's outlet from Rhodesia. He obtained the railway concession that the Portuguese government had made available and in 1903 he founded the CFB with 80% British capital.

Paulings was given the construction contract and started work in March 1903 on the new 1 067 mm-gauge line; but they twice had to stop when they ran out of money, and in 1904 Norton Griffiths contracted to find the finance and do the work on a cost plus basis. According to George Pauling, this system proved so costly to the CFB that in 1909, when Griffiths had completed the first section up to kilometre 150 and had brought the rails as far as Cubal at kilometre 197, Paulings, financed by Erlangers, was again given the contract and all subsequent ones until the Congo border was reached in 1928.

Building the Benguela Railway was one of those feats of human endurance and perseverance so common in the history of railways in Africa. The first 150 km were the worst because they ran through the inhospitable and unhealthy 'thirsty country' and because the contractors were forced to build in a rush along an unnecessarily difficult route. So much time had passed between the granting of the concession and the signing of the contract by Griffiths that they were left with only ten months to reach Catengue at Km 122, if the terms of the concession were to be met – and they knew a German firm was eager to take over the concession should they default.

Many of the contractors' difficulties could have been avoided had the concession not stipulated that the line should follow the route surveyed by the Portuguese when Caconda was one of the main objectives. A shorter, easier and more direct route was eventually followed with the building of the Cubal variant in 1975; if this route had been followed from the beginning it would not have been necessary to build the rack section or its 1948 replacement.

Varian, the engineer, describes the building of the rack section, where 'every possible form of transport was pressed into service over the first three-thousand-feet summit, where wheeled transport was of no practical use'. Camels and human porters were brought in, as were rock drillers from Cape Town 'to hew the bed on which the rack was laid out of the solid granite of the steep mountainsides'. Electric lights were strung up to allow continuous shift working. Here, and on the whole section before Cubal, 'every conceivable form of tropical sickness was encountered' and many died; all water had to be

196

195 (Previous page). **Train time at Golungo Alto, terminus of the winding 600 mm branch from Canhoca. The daily mixed, pulled by a 1931 Henschel 0-8-2T, one of three active steamers on the line in 1970, is met by crowds reminiscent of an age before motorcars and aircraft.**

196. **With distinctive Benguela Railway features, these two engines of the CFB 11th Class differ externally from their SAR 19D counterparts in having illuminated number boards on headlamps, lugs on smokebox doors, and long-pointed cowcatchers.**

carried by train from Benguela; and 9 000 labourers, mainly Indians from Natal, and West Africans, had to be brought in.

After the initial lack of finance, delays in construction were caused by World War I, and by the shortage of materials in the post-war period. Then, just as the end was in sight, the Congo border was moved, after Belgium exchanged land in this area for some Portuguese territory at the mouth of the Congo – the Benguela Railway had an extra 106 km to go before it could relax and wait for the connection on the Katangan side to be completed. This accounted for a further delay, and the link was made only in 1931, by which time much of the copper traffic was committed to other routes, particularly the *voie nationale* of the Belgians. In spite of the fact that this was inefficient, involving several transhipments from rail to river, it was only after World War II that the CFB began to take 20% to 40% of Katangan copper.

A motley collection of construction engines preceded the first road locomotives of the CFB, classes 5, 6, 7 and 8, introduced from 1908. Class 5 comprised the Esslingen-built rack engines which for 40 years plied the rack section between Chivanda and São Pedro. When this section was replaced in 1948 by the 1-in-40 Rio Lengue route, class 5 locomotives, with rack gear removed, were used for shunting, surviving into the 1970s. Classes 6 (4-6-0), 7(4-8-0) and 8(4-8-0) were built to existing Cape Government Railways designs for their classes 6, 7 and 10 respectively. Of the original main-line engines, only class 6 stayed on the CFB, classes 7 and 8 being sold in the mid-1950s to the Moçamedes Railway further south.

The original Cape Government Railway class 10 had been a single experimental machine; but as CFB class 8 the design was well suited to Angolan conditions, which called for a large, shallow grate for burning wood, and low drivers for plugging away at the long grades up from the coast. Development of this type continued through classes 9A and 9B, culminating in the ultimate Angolan 12-wheeler, the North British 9Cs of 1930, which all survived in road service into the 1970s.

The limitations of non-articulated power on the tremendous haul up the escarpment led CFB to become the first to use double Mountain Garratts; its

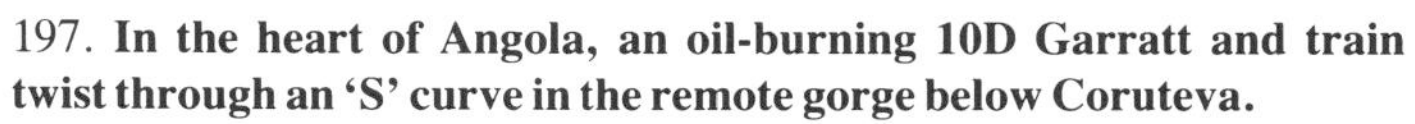

197. **In the heart of Angola, an oil-burning 10D Garratt and train twist through an 'S' curve in the remote gorge below Coruteva.**
198. **A contrast to latter-day Garratts: No. 01 started its life as a construction locomotive and ended it as a 'relic', parked under palm trees at Lobito station.**
199. **This former 0-6-2T rack locomotive, No. 14, was relegated to shunting services after a distinguished 40-year career on the old rack section up the Lengue Gorge, replaced in 1948 by a new line.**
200. **Benguela 6th Class No. 26 is a slightly 'modified' version of the South African Railways 6th Class, several of which were sold to Angola between 1907 and 1910.**
201. **CFBs first Garratts were not only the first double-mountain Garratts in the world, but unique in that they were equipped with Lentz poppet valves, clearly visible on No. 302, at Nova Lisboa (now Huambo). The sparkling appearance of this engine is typical of the pride of the management and staff of the CFB, who maintained their steam power to standards as high as could be found anywhere.**

197 198
199 200
201

202. **A 'Dupla' in Portuguese terminology, these two oil-burning 10D Class Garratts pour on the smoke with a Benguela-bound load of ingots from the Zambian copperbelt.**
203. **A misty, drizzly morning at Coruteva, midway on the climb from the coast at Benguela to Cubal. Two Garratts – the foremost oil-burning, and the second a wood-burner – heave mightily to start a combined train on slippery rails.**

class 10, a Beyer Peacock design with Walschaerts-operated Lenz poppet valves and plate frames, was introduced in 1927. So successful was this class that from 1927 Beyer Peacock had the CFB locomotive market cornered, applying bar frames and piston valves and producing its 1927 design, with detail modifications, as classes 10B, 10C and 10D right up to 1956. A further nine Beyer Peacock engines were bought from Rhodesian Railways in 1964; the RR class 16s were reconditioned and placed in service as class 10E.

After 1930, the only non-articulated (and only non-Beyer Peacock) power the CFB acquired was six 4-8-2s of class 11, in 1951. These engines were practically identical to the class 19D, but the addition of CFB details made them more attractive than their counterparts down south.

At CFB's main workshops at Nova Lisboa (Huambo), surprisingly heavy work was carried out, and the ingenuity of the mechanical engineers increased in direct proportion to the decline in the number of suppliers of steam locomotive parts. Thousands of patterns for various castings were held in stock, and CFB took special pride in its own patterns for self-adjusting pivots, equipping all 10, 10B and 10E locomotives with them as they came into shops. When introduced by Beyer Peacock in the late 1940s, these pivots were one of the most important innovations ever applied to Garratts. Heavy boiler repairs were also undertaken at Nova Lisboa, the wood-burning locomotives needing far less maintenance than the oil burners did.

The CFB was one of the few railways to continue to rely almost totally on steam traction into the 1970s, thanks mainly to the Garratt design, which extended the useful life of steam locomotion almost everywhere it was implemented.

In 1971 construction began on the 170-km deviation known as the Cubal Variant which runs from Catumbela to near Cubal. This undertaking, a huge one for a private company, took five years to complete and reduced the grade up from the coast from 1-in-40 to 1-in-80. The first CFB diesels were ordered for this new, water-short line. At the same time it was decided to replace steam between Lobito and Nova Lisboa, eliminating Garratts from the entire escarpment section. Following the Portuguese withdrawal from Angola, however, events overtook this programme and it is uncertain whether any steam remains in use here.

As a result of the civil war the CFB has not operated as a through railway since 1976, and is no longer a major outlet for copper and other minerals from Zaire and Zambia. This traffic now passes through South African ports, with the new Tazara Railway taking the overflow to Dar es Salaam.

Much traffic was also generated along the route, notably timber and maize, and iron ore on the one short branch, from Robert Williams to Ciamo. Manufactured items and fuel formed the bulk of the traffic moving inland.

Lobito was connected by rail with Cape Town, 6 000 km away, and until well into the 1970s it was possible to cover 4 000 km of this journey behind steam locomotives on scheduled trains. One could ride the CFB section in a mahogany-panelled diner, the smells from the eucalyptus-burning stove and the eucalyptus-burning Garratt up front mingling in one's nostrils.

From the passenger's viewpoint and in every other way, the CFB was undoubtedly one of the most smoothly-run rail operations in Africa – from the neatly trimmed ballast of its roadbed to the spotless conditions in the backshops, where never a lead plug seemed out of place.

Caminho de Ferro de Luanda (CFL)

Travelling along the main highway from Nova Lisboa to Luanda in September 1970, soon after descending the scenic Angolan escarpment, we came upon the little town of Dondo. There, in a decrepit old engine house, we found a shining 1924-vintage Henschel 4-8-0, still warm from its morning trek with the daily mixed from Zenza. The branch from Zenza to Dondo was a good introduction to the northernmost common carrier in Angola, the Caminho de Ferro de Luanda (CFL), for the CFL is the oldest line in Angola and the Dondo branch still had the unsophisticated charm of a pioneer African railway, with its unballasted track, light rail, metal sleepers and minimal earthworks.

Building was started from Luanda in 1886 by the Cia de Caminho de ferro atravez d'Africa (Trans-Africa Railway Company), but Rhodes thwarted its ultimate aim when he colonized Rhodesia in 1890, so cutting off Portuguese West Africa from Portuguese East Africa. Thus the CFL was relegated to a 400 km penetration line. The company completed the first section to Ambaca at Km 340 in 1899, the Portuguese government having guaranteed interest on its investment. It then tried to continue to Malange without government help, but ran out of funds and the government took over construction reaching Malange, its present terminus, in 1909. In 1914 the state took over the entire line, converting it from metre to Cape gauge in 1963.

Except for the 250 series 2-8-2s, all the still serviceable CFL steam locomotives were converted in 1963, since when CFL power was freely interchangeable with that of the Moçamedes section of the CFA and examples of most classes could be found on both systems.

In 1970 the CFL roster listed 26 active steam engines, all of which were oil-fired, with another seven stored but serviceable. The bulk of road freight work was done by steam, but four General Motors diesels handled the mail and mixed train schedules.

An assortment of light continental tank locomotives, dating back to the start of construction, preceded the first business-like power acquired by CFL. These were the 150-series Henschel and 200-series Armstrong Whitworth 4-8-0s of 1924. The British firm also supplied an ungainly low-drivered 4-6-0, presumably for branch-line service. These 50 or so engines bore the brunt of road service until 1949, when Beyer Peacock sold six of its worthy products to the line.

Ideal for sharp curves, heavy gradients and light rail, the Garratts were deservedly popular, and in 1954 when further power was needed Garratts were again the choice. This time the order went to Krupp, who set about designing and building a heavier and more powerful machine than the Beyer Garratts. Whereas the British firm never seemed concerned about the aesthetic qualities of its products – most being strictly utilitarian in appearance – Krupp made a real attempt at styling, which resulted in some striking machines.

The German Garratts, although well-engineered, were not as popular as

205

206

207

208

the pedigreed products of Gorton. Krupp's wheels were smaller and they had a reputation for rough riding; nor were the Krupps as economical in terms of fuel consumption; and their maintenance costs were higher, mainly as a result of their low drivers and of Krupp's inability to use Beyer Peacock's patented self-adjusting pivots in their design.

In the mid-1950s the Portuguese government planned a brand new 'Congo Railway' to run northward from Luanda towards Cabinda. This line was to be Cape gauge and at the same time the decision was taken to widen the existing CFL lines.

Motive power for the new line took the form of ten 2-8-2s built by Arnhold Jung in 1956. These beautifully constructed machines, numbered in the 250 series, were sent initially to the Moçamedes Railway (CFM), pending completion of the northern line. However, early in the 1960s, after Congo independence, construction was suspended after only 10 km had been built and, only after the 1963 gauge conversion, could employment be found for three of the Jungs on CFL; the others remained on CFM.

Principal traffic on the Luanda system consisted of coastbound iron and manganese ore and agricultural produce, notably coffee and maize, while manufactured goods and fuel were carried inland from the coast. Non-articulated helpers were used on Garratt-hauled trains out of the Luinha River valley. The helper, usually a 200-series 4-8-0, came on at Canhoca and both eastbound and westbound freights were assisted where necessary.

From Canhoca eastbound, the line climbs the escarpment to Salazar (Dala Tando) and relocations resulted in heavy earthworks along this scenic stretch. Once a train reached the plateau beyond Salazar the heavy grades were over and only slight undulations remained before reaching the terminus at Malange.

Angolan maps show a proposed eastward extension of the CFL, which eventually turns southward and connects with the CFB. Whether such a line will ever be built is uncertain, but it would open vast tracts of almost undeveloped country. More likely, is the projected line, reaching northward from Luanda towards the Congo River and for which earthworks for the initial 45 km are completed.

From Canhoca, 30 km west of Salazar, a 600 mm-gauge branch ran north for 30 km up to the coffee plantations around Golungo Alto. Built in the 1920s, this narrow gauge was a quaint operation, with tiny wood-burning tank locomotives and bright red, clerestory-roofed passenger coaches with wrought-iron hand railings around the platforms at each end. The daily mixed train was well patronised – probably because of the incredibly bad state of the adjacent road. The line passed through tropical rain forest with wayside stations which were rural in the extreme.

Caminho de Ferro do Amboim

The CFA's Caminho de Ferro do Amboim, a 130 km-long 600 mm-gauge railway, was completed in 1925 to serve the extensive coffee plantations on the Amboim plateau around the town of Gabela at an altitude of 1 100 m. In 1970 all shunting around Amboim was done by small diesel-mechanical putt-putt machines, and a railcar ran twice a week, but otherwise the railway was entirely steam-driven. Six wood-burning tank locomotives maintained an uncertain service of one 'misto' (mixed) and two freight trains each week. Every train operated meant at least two days away from home for the locomotive and crew, because it took an incredible 12 hours to cover the line in each direction.

The long-suffering steam locomotives on the misto had a relatively easy time for the first 80 km out of Porto Amboim, headquarters of the line. After Boa Viagem (Carloango), however, the fireworks began, with a triple horseshoe below Lacetes, after which the line clung to a granite shelf above airy drops, before reaching a mid-section water-stop in tropical rain forest. Thereafter, punctuated by numerous unscheduled stops to blow up steam, climbing continued through sweet-smelling coffee plantations until, with a triumphant gasp, the misto ground up the main street of Gabela towards its depot. Its arrival was greeted by the villagers like the return of a long lost friend, and the last stretch was accompanied by joyful shouts of 'El comboio' from the local children, who swarmed over the slow-moving train for a free ride to the depot a few hundred metres away.

Amboim coffee beans are highly sought after for the manufacture of instant coffee, but the line's yearly freight seldom surpassed 20 000 tonnes and there were only 12 000 fare-paying passengers annually. It is sad to record that it closed in October 1975.

Caminho de Ferro de Moçamedes (CFM)

For many years the southernmost of the Angola railways, the CFA's Moçamedes system, or Caminho de Ferro de Moçamedes, was of little importance other than to the immediate communities it served. It was built by the Portuguese government from 1905, proceeding gradually as loans to finance it became available, and reaching Lubango (Sa da Bandeira) only in 1923. For a long time the terminus remained at Huila (Vila da Ponte), 249 km from Moçamedes. The line was built on 600 mm gauge, trackwork was of the lightest, and CFM's motive power was provided by Orenstein and Koppel 0-8-2 tanks, assisted from 1929 by three ex-SAR class NG9 Baldwin 4-6-0s; the only reason for its existence was the impossible nature of alternative communications.

In the 1950s, with aid from Portuguese government development funds, the gauge was widened and the main line was extended, reaching Serpa Pinto (Menongue), about 800 km from Moçamedes, in 1958. At the same time, hand-me-down 1 067 mm power was acquired from the CFB, in the form of their classes 7 and 8 and later six beautiful new Garratts were delivered by Henschel.

Large quantities of high-grade iron ore had been discovered south of the line around Cassinga and traffic was outgrowing the capacity of even the big Henschels and the 10 Jung 2-8-2s on loan from CFL. In 1967, when the Krupp-financed branch to Cassinga was completed, 45 new diesels were ordered which virtually put paid to steam power and by 1970 most of it was in storage.

210

204. (Previous page 204-208). **Departure of the 17h00 Lobito-Silva Porto train in August 1972, with 4-8-2, No. 401, providing the steam and smoke effects.**
205. **At the same location but with a lesser train, the 14h19 Lobito-Benguela local, headed by a Baldwin 4-8-0, No. 215, prepares to depart.**
206. **On the final climb from the coast to the inland plateau, wood-burning Garratt, No. 303, climbs through a succession of horseshoes above Lepi.**
207. **The CFM, or Moçamedes line, was the first in Angola to put steam to pasture. Among the casualties was this large Henschel Garratt, No. 102, one of a series of six engines built in 1953.**

208. **Pretty as a picture – a 'Princess' Class 2-8-2 rests on the turntable at the Luanda roundhouse in 1970.**

209. **Speed is not an African characteristic, particularly on an unimportant CFA branch line. Nevertheless, 4-8-0 No. 156 gets a roll on, working the mixed from Dondo to Zenza.**
210. **Krupp built CFAs five final Garratts in 1954. Few other Garratts had so unique an appearance as these, which included No. 555 – seen leaving Luanda during 1970.**

211

211. The Porto Amboin-Gabela narrow-gauge was an operating gem. Described as 'the Darjeeling of Africa', it made a memorable experience of any journey on its trains. But the thoughts of the local passengers, boarding the once-a-week 'Misto' at Gabela, were probably less of aesthetics and more of the slow and tedious journey ahead.

212. In the heat of the coffee plantation a downhill train rumbles through a green canopy of forest.

213. Mist often shrouds the escarpment, giving the forest a special magic. Here 2-8-2T No. 40 cautiously works over switch points into an isolated siding. This engine was one of two 1942 Bagnall-built 2-8-2Ts (Nos 40 and 41), the regular power for the 'Mistos'. For freight traffic two chunky 1949 Henschel 2-8-0Ts (Nos. 60 and 61) were the usual power, while three older Henschel 2-6-0Ts dating from the 1920s were on the property although only two were serviceable in 1970. Finally, a 1921 Linke-Hoffman 0-6-0T, possibly used during construction days, occupied an unused corner outside the Porto Amboim shed. CFAs newer locomotives carried the following names: 40 'Amboim', 41 'Cuanza Sul', 60 'Ebo' and 61 'Boa Entrada'.

212

702
C F M

9 EASTERN LINES

MOÇAMBIQUE

Moçambique owes its intriguingly varied railway system as much to its colonial past as its excellent natural harbours. These two factors have led to the country developing six entirely separate railway systems, reaching inland from the coast to Zimbabwe (and then to Zambia and Zaire), to South Africa, to Swaziland and to Malawi.

But when the authors set off in 1969 on an 8 000 km round journey, the attraction was not the multi-faceted nature of Moçambique's railways, but four particular locomotives – Atlantics believed to be among the last of their kind operating anywhere.

'We had heard that these engines were still in service, operating from the northern town of Nampula', they recall. 'Sixteen days later we had proof, not only of the Atlantics but also of many other steam locomotives of the Caminhos de Ferro de Moçambique (CFM). We had seen some 150 engines of 35 classes and, considering CFMs roster of 237 engines of 40 classes, the outcome of the trip was satisfactory. Just as Spain had been to enthusiasts ten years earlier, so was Moçambique in 1969, hardly a day passing without some new and exciting discovery.'

Since independence, reports indicate that even if the railways are still largely intact, there has been a dramatic drop in traffic. However, the Maputo (formerly Lourenço Marques) – Komatipoort section, operated with the assistance of the SAR, is as busy as ever.

Rail development in Moçambique dates from the 1880s when the Delagoa Bay Railway was constructed inland from Lourenço Marques towards the Transvaal Republic border as part of a scheme to give the young Boer republic access, independent of British influence, to the sea.

This set a precedent, for Moçambiqué's first railway was, as a result, not constructed primarily for her own internal needs, but as a bridge route for a neighbouring country. With two major ports – Beira and Lourenço Marques – as well as several lesser ones, Moçambique lent herself to the role of sea link for the landlocked countries west of her. By comparison, Angola has only one international route; however, neither of these former Portuguese colonies has a north-south internal through-route. This has led to railways with the individual character that isolation breeds, although in the case of Moçambique all systems are state-owned and have been controlled, with two major exceptions, by a largely centralized authority since the late 1920s when the Carta Organcia was established in Lourenço Marques.

The major railways traverse the country from east to west and major development has tended to follow the direction of the line of rail. Therefore in the past there has been little demand for an internal north-south service and the existing railways have effectively divided the country into three largely independent areas. In theory, however, it is possible to travel between these three major systems (the Lourenço Marques, the Beira and the Moçambique) by passing through one or more adjacent territories and taking advantage of Moçambique's numerous international rail links. This is hardly practical. For instance, the rail distance from Lourenço Marques in the south to Nacala in the north would entail a trip of more than 3 200 km, whereas by air it is less than half this distance. Traditionally, movement north and south has been by sea, and it was only in the early 1970s that the first tar road running north from Lourenço Marques to Beira was completed.

214. **CFMs finest, a 1948 Montreal 4-8-2, emerges from Alto tunnel on the Swaziland railway line to Ka Dake. These locomotives, built first for the Lourenço Marques-South Africa (Ressano Garcia) run, later spent seven years in Swaziland, hauling ore on the upper section of this scenic railway. Similar in size to a South African Railways 15F 4-8-2, these engines had detail refinements such as cast-steel beds and a distinctly American appearance, pleasing to most who saw them.**

When the Carta Organcia came into being the state controlled some 60% of the existing lines (less than 1 360 km), the remainder being split between the Beira Railway, operated from Rhodesia, and the Trans-Zambesia Railway, a British-owned company.

In the southern part of the country, the government operated the busy Lourenço Marques-Komatipoort line, including two branches – one to Goba and the other to Xinavane. A 750 mm-gauge line running north from Lourenço Marques to Marracuene, 35 km distant, was later converted to 1 067 mm and is today part of the route from Lourenço Marques to the Zimbabwe border at Malvernia. Farther north, but still in the southern part of the country, another 750-mm line ran 86 km inland from Vila Nova de Gaza (later renamed João Belo) to Chiomo. Farther north still, another isolated line, this one 1 067 mm, ran parallel to the coast between Inhambane and Inharrime. Today, a tar road runs parallel to this line, competing with the already under-employed railway.

The Portuguese did not control the two railways operating from Beira and the remaining state-controlled railways lay farther north. First was another 750-mm line, operating from the small port of Quelemane. In later years, this was converted to 1 067 mm. Finally, at a latitude of 15° 4′ S, and well within the tropics, a 1 067 mm-gauge railway ran inland from the obscure port of Lumbo. It was for this line that Henschel, in 1923, built the four Atlantic 4-4-2 engines which were in later years to attract so much international attention from enthusiasts. These engines have spent most of their lives on this line, although at first they were diverted to the Lourenço Marques system where they operated until 1933.

By 1969 the CFM had grown into a 3 624 route-kilometre system, giving it 30% more railway than Angola, although its area is 35% smaller. The consolidation of the CFM had come in stages, first with the acquisition of the Beira Railway from the Rhodesians in 1949, and the take-over of the Trans-Zambesia Railway (TZR) in 1969. New construction after 1929 included the gradual extension inland of the Lumbo-Nampula line to a terminus near the shores of Lake Malawi at Vila Cabral, and finally the linking up to the Malawi State Railways north of Blantyre, in 1970.

Further south, in the Zambesi Valley, another line was built running from a junction on the TZR at Sena to the rich coal fields on the north bank of the Zambesi at Moatize. This line, opened in 1949, later served as an important supply route for the construction of the Cabora Bassa Dam and was the focus of much guerilla activity until independence in 1974.

From Lourenço Marques two new international routes were opened: the first in 1955, ran to the Rhodesian border at Malvernia, and the second, to the Swaziland border beyond Goba, opened in 1964. The latter extension was part of the Swaziland Railway project, which included a further 219 km in Swaziland to an iron ore mine at Ka Dake. Finally, a less important branch was constructed to the south of Maputo, terminating at Salamanga.

All this development increased CFM's locomotive requirements. From humble beginnings in 1887 when its stable consisted of four locomotives (two 4-4-0Ts and two 4-6-0Ts), some 300 engines of 28 wheel arrangements and built in seven countries were eventually obtained. More than a third of these were acquired second-hand, and some were even third-hand.

As the system grew, so the variety of locomotives increased. At least 50 classes, built by more than 20 locomotive manufacturers, have operated on CFM rails – and there are few systems of this size with a comparable array. Admittedly, this tally includes 30 units of one design, which indicates some degree of standardization, nevertheless, Moçambique offered a marvellous diversity of steam.

An inspection of CFM's roster shows the development of distinct buying trends, based on a preference at any given time for builders from particular countries. The 1 067 mm-gauge roster reveals that in 1912, after 24 years of new and second-hand locomotives built in England and Germany, the first American engines were acquired. These were two large 2-6-6-0 Mallets, similar to those in operation on the SAR. The reasons for buying such monsters are obscure as there were no typical mountain climbing lines on the Maputo system at the time. Over the next 12 years a further 19 American 2-8-2s, 2-10-2s and 4-6-2s were acquired, and these set the standard for basic engine dimensions that were followed for new non-articulated mainline power until the 1950s.

C F M

C F M

215 216

217 218

215. **Atlantic over the Indian Ocean: No. 814 at Lumbo in 1969.**

216. **Stunning — is the word for this ex-works Henschel Garratt at Beira.**

217. **A Belgian-built Garratt, of British concept and patent, in a Portuguese-speaking country — and an African setting.**

218. **Contrast to the country. Baldwin Pacific No. 302 leaves Lourenço Marques station whose copper cupola dominates the urban skyline.**

In 1922, a 'continental' period began with the delivery of three small 2-8-0s of Gallic design for the Inhambane line. These were followed by Henschel's four famous Atlantics, and six 'brother' 2-8-2s in 1925. These ten locomotives were built for the Nampula line, but operated for a time from Lourenço Marques – giving the authorities a chance to compare the German products with the Yankee engines.

No more engines were acquired until 1930, when the first of three 2-8-0s for the newly widened Quelemane line arrived. The same year saw the delivery of an isolated O. & K. 2-8-0 for Inhambane, and this was followed by two new classes – nine 0-10-0T shunters and six lightweight 2-6-2s. Orders for German locomotives might have continued but for the advent of World War II. Instead, the Moçambique authorities had to find a new source of motive power and the Americans were eager to oblige. H.K. Porter delivered two American versions of the German 2-6-2s in 1941. This was followed by three additional 2-10-2s of the original 1915 type in 1945. After the war, German industry could not supply locomotives, and the CFM continued to buy American products – first six Baldwin 2-8-2Ts, then ten 2-8-2s almost identical to the four delivered 24 years earlier. During this 'American' period, Canada, too, built eight huge Mountain type 4-8-2s in 1948. These impressive machines are considered by many to be CFM's finest steam engines.

The acquisition of the Beira Railway gave CFM first-hand knowledge of Garratts, as they took over 17 former Rhodesian locomotives as well as the railway. With traffic increasing it seemed logical to acquire more of the same and within three years the Beira division began taking delivery of 12 large double Mountain Garratts from Haine St. Pierre. At the same time four former SAR class GF Garratts were obtained, mainly to work light passenger trains. These were followed in 1956 by five Henschel-built 4-8-2+2-8-4s, the largest Garratts to run on the CFM and the most distinctive engines in Moçambique. Contrary to past practices, these engines were not painted black, but instead were finished with maroon tanks with gold trim, light gray boilers, and fire-engine red wheels with white rims. These were among the most beautiful Garratts ever to have turned a wheel on any railway. CFM's final Garratts arrived in the 1960s. Three of these were British WD 2-8-2+2-8-2s, of the same type as those acquired when the Beira Division was taken over in 1949, which came from the Congo Ocean Railway. They were followed, in 1964, by even more interesting locomotives – ten 4-6-4+4-6-4 Garratts, built in 1936 for the Sudan, and sold to the Rhodesian Railways in 1949. These handsome machines brought to 51 the total of Garratts used on the CFM.

In the south, at Lourenço Marques other, different developments were taking place. There was, in effect, a German renaissance, beginning with the purchase of eight harbour shunters – 0-8-2Ts from Henschel – in 1950. These were followed in 1951 by six Henschel 2-10-2s, updated and more modern versions, with larger boilers than the 36-year-old Baldwin design.

There followed a lull until 1955, when CFM's largest ever steam locomotive order was placed – again with Henschel. It comprised 33 engines of five types: two 0-8-2Ts, four 2-8-2Ts, two 2-8-2s, three 4-6-2s, and 22 2-10-2s. To stable these additional locomotives, a large new roundhouse was built – a welcome change from the usual parallel-road sheds so much a part of the scene in recent times at loco depots in southern Africa.

The delivery of these engines to Lourenço Marques and the Garratts to Beira in 1956, marked the end of CFM's acquisition of new steam locomotives. The following year the first diesels – Krauss-Maffei hydraulics – made their appearance and it seemed that CFM had turned its back on steam.

But although no new locomotives were bought, 66 more second-hand steam locos were acquired between 1956 and 1976 from three African countries. The French Congo (now Congo) supplied the first of these – eight 2-8-2s and later three WD 2-8-2+2-8-2 Garratts, similar to those which came with the Beira Railway take-over. From 1962, Rhodesia Railways sold 30 Mountains of four classes to the CFM, as well as the ten Sudanese Garratts mentioned earlier. Finally, SAR sold two classes of 4-8-2s: three 15Es in 1971, and 12 15BRs in 1972-73. The purchase of these engines earned CFM the distinction of having purchased new steam locomotives over a 69-year span, and second-hand steam locos over a 76-year period. There are few railways with such a record.

Developments in Moçambique are affected by the current political situation. When the border with Rhodesia was closed in 1976, there was a dramatic drop in traffic on the Beira line and until recently the Maputo-Malvernia section was little more than a streak of rust. Throughout the rest of the country economic activity is so depressed that railway traffic probably has dropped. SAR has helped maintain the flow of traffic to and from Maputo – particularly to the border with South Africa – and has leased both diesel and steam locomotives to CFM. In 1978-79 a batch of 25 Brazilian General-Electric diesels arrived in Moçambique, supplementing the 48 existing diesels and a recent order for 25 more diesels could see the end of steam operation on the Maputo section. But events in Moçambique are unpredictable.

The reopening of the border with Zimbabwe has heralded a return of traffic to the Beira line, but CFM has announced that electrification of this section will begin late in 1980. Thus the prospect of a major resurgence of steam coinciding with the country's economic recovery has been dealt a mortal blow – though in the short-term steam could still play an important role. If it does, we hope to be on hand to record it.

SWAZILAND

Though Swaziland is one of Africa's six smallest mainland countries its railway operation was anything but small. Its 215 km mountain railway reached an altitude of 1 500 m on long 1-in-50 grades and featured double-headed former Rhodesian 4-8-2s, sporting CFM numberplates.

The railway was opened in November 1964, making it one of the youngest on the continent and was built to serve an iron ore mine in the western part of the then Crown Colony, which gained independence in 1968. Although the railway owned its wagons, locomotives were supplied on hire from Moçambique. This was convenient as ore had to move down to Lourenço Marques (now Maputo) through the border at Goba, and CFM had surplus steam power available, while Swaziland had coal and good water. Diesels were not considered an economic proposition and part of the line remains steam operated.

However, Moçambique diesels operate from Maputo into Swaziland at Sidvokodvo, the 'half-way' engine change point, while beyond to the mine at Ka Dake, South African 14th Class 4-8-2s do the work. These unglamorous plodders took over early in 1978 when Moçambique called its large, Montreal-built 4-8-2s back home. The mine was to close in 1980 and the sound of heavy exhausts high above the Usuto River may have passed into history.

But this is not the end of the Swazi steam story for, in 1978, a new 93 km railway was opened reaching from Phuzumoya, down through Big Bend to the SAR Natal system at Gollel and steam power is active on this line. General cargo is increasing and the industrial area at Matsapa is growing, requiring two trains a day from Sidvokodvo. A plan to connect with the SAR at Komatipoort will allow through running of trains from the Eastern Transvaal to Richards Bay in Natal – which could bring even more traffic to Swaziland's railway.

219. **The Henschel Garratts were the largest locomotives to run in Moçambique. Their sphere of operation was normally limited to the Vila Machado–Gondola–Machipanda portion of the Beira main line, where they worked both passenger and freight trains. In 1975, near the Rhodesian border, a scruffy 970 works past a blockhouse, relic of the recent conflict.**

220 (Following page). **Late in the afternoon, in the heart of the Amatongas forest, a 970 Class Garratt works a coastwise goods up a short incline. When first constructed, as a 2′0″ (600 mm) gauge, Beira Railway included reverses through this hilly area.**

221. Moçambique Railway locomotives monopolized operations on Swaziland Railway from its opening in 1964 until early 1978, when South African Railways began to supply motive power. During the transition period a CFM 700 Class 4-8-2 leads a SAR 14R 4-8-2.

222. In the early 1970s CFM equipped No. 707 with a Geisl ejector, but did little proper testing. Nevertheless, this engine successfully operated, seeing regular service in Swaziland.

223. Not all traffic in Swaziland was iron-ore, and a short branch to the growing industrial area of Matsapa, near Manzini, has seen greatly increased traffic. Two former Rhodesian 4-8-2s – of the type which first operated ore trains, prior to the arrival of the 700s – haul a load shortly after sunrise.

224. Summer in Swaziland and rich green vegetation contrasts with the dull red of empty ore trucks. The upper section to Ka Dake was scenically spectacular in places, and of great interest were the few miles south of Nondvo, where three tunnels, deep cuttings and numerous sweeping curves marked the route.

221

222

223

224

225

226

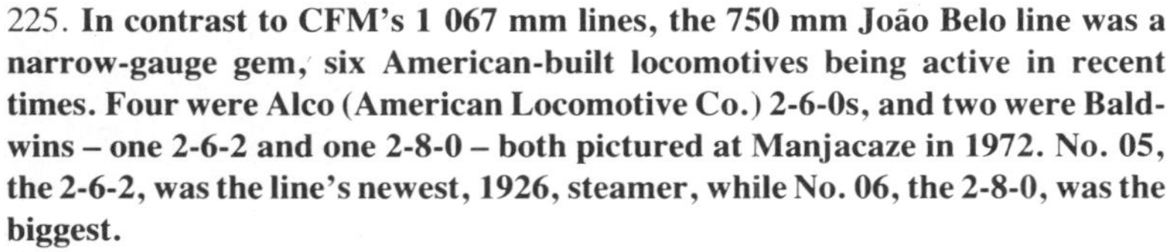

225. In contrast to CFM's 1 067 mm lines, the 750 mm João Belo line was a narrow-gauge gem, six American-built locomotives being active in recent times. Four were Alco (American Locomotive Co.) 2-6-0s, and two were Baldwins – one 2-6-2 and one 2-8-0 – both pictured at Manjacaze in 1972. No. 05, the 2-6-2, was the line's newest, 1926, steamer, while No. 06, the 2-8-0, was the biggest.
226. No. 05 was originally built as a 0-6-2 tender engine, although rebuilt to 2-6-2 in the early 1970s. In her original state she was photographed at Manjacaze in January 1970.

227. A typical rural scene along the line, with the Saturday mixed for Manjacaze to João Belo, with 0-6-2 No. 05.
228. An almost timeless scene as No. 05 leaves Marao on Christmas Day 1972. This 140 km line was at last report 'still going' though two changes had been made, slightly modernizing the railway – fuel was switched from wood to coal, and the engines were equipped with sealed-beam headlamps.
229. Cleaning the ash from the smokebox – a job which would be rendered unnecessary if a modern self-cleaning front end were fitted.
230. Before the change to coal, wood sparks burned holes in clothing and sometimes even the wearer's skin.

227

228 229

230

CAMINHOS DE FERRO
8
MOCAMBIQUE

231. The architectural charm of the engine shed at Inhambane is as delightful as it is unusual. From one of the arches No. 8, a St. Léonard-built 1 067 mm-gauge engine, and one of five active steam locomotives on this isolated 90 km line, views the workday ahead.

232. This Gallic treasure, No. 6, is one of three 1922 St. Léonard 2-8-0s operated by CFM on the Inhambane line. Here she hauls a southbound passenger train – the first carriage of which is itself of classic proportions.

232

233

233. Train time at Jangamo – 34 km south of Inhambane. In spite of a parallel tarred highway nearby, trains were well patronized until Moçambique's independence. Today, little is known of the state of this line and its locomotives.

234. Through groves of date palm trees – a tropical setting in the best tradition – comes Baldwin 2-8-2 No. 406 with the Quelemane-Mocuba mixed, in July 1969. No. 406 is one of 10 second-order 1948 Baldwins (405-414), virtually identical to CFM's first American 2-8-2s (No. 401-404, Baldwin, 1924) although equipped with utilitarian stovepipe chimneys. The Quelemane line was originally part of a scheme to construct a railway on the north side of the Zambezi to Nyasaland, although it only reached Mocuba, 159 km inland, leaving another 140 km to the border.

234

235

235. **After steam is gone some alternative 'watering arrangement' will have to be made for the local population at Gondola.**

236. **Cab-side plate of No. 971, after a major overhaul in December 1972.**

237. **Last of the ZASM Bs. CFM No. 42, pictured at Rio Monapo in 1972, was one of more than 30 such former Transvaal (NZASM) locomotives acquired over a ten-year period from the late 1890s. Surprisingly, four survived into the 1970s, all the more remarkable since of more than 200 built, the first went to scrap before 1910.**

238. **Busy scene at Lourenço Marques (now Maputo), as a Porter 2-6-2, No. 572, approaches with the 05h30 train from Tenga, while a Baldwin 2-8-2T shunts on the adjacent tracks.**

239. **Porter No. 571 was transferred to Inhambane during the early 1970s, supplementing the line's three St. Léonard and one O&K 2-8-0s. In March 1975, on the morning train from Inhambane to Inhirrime, she sits at Jangamo, lovingly cleaned by her regular driver.**

CAMINHOS DE FERRO
971
MOÇAMBIQUE
P.R.O.G. 6-12-72
HENSCHEL
— 1956 —
HENSCHEL & SOHN KASSEL

C.F.M.
42

C F M

238 239

C F M

241

240. **German immigrants to former Portuguese Africa. Sena Sugar Estates operated extensive 'tramways' near the mouth of the Zambezi and of two dozen 600 mm-gauge steam engines on the roster in 1969, 14 were old 'Feldbahn' 0-8-0Ts.**

241. **Sena Sugar Estates also operated a 100 km 3′0″ (914 mm) gauge main line, which featured 11 locomotives of four designs. At the shed (from left to right) are a 1954 Peckett 0-6-0T, two 2-6-4Ts, and an 0-6-2T. The last locomotive built by Peckett, and shipped from Bristol on June 12, 1958, became Sena No. 7, and was seen in steam in July 1969.**

242. **One 2-10-2 is fine, two are better. During the early 1970s this was a familiar sight in Swaziland, where Moçambique CFM 2-10-2s operated ore trains from Sidvokadvo – midway on Swaziland's railway – down to Lourenço Marques. The Lebombo Mountains provide a backdrop at Mlawula, near the Moçambique border.**

242

MALAWI

'The Gateway to Central Africa is the Zambezi'
David Livingstone

243

Malawi's railways are so closely linked to those of Moçambique as to seem a mere extension of the former Portuguese network. And though in earlier days they were an extension of British colonial railway interests in Moçambique, nothing could be further from the truth as they have a character very much their own.

The railway into Nyasaland (as Malawi was known until independence in 1964) was conceived as a feeder from the Shire River, a tributary of the Zambezi. Before the railway was built, access to the healthy highlands adjoining Lake Nyasa had been by stern-wheel paddle-steamer from the mouth of the Zambezi at Chinde, upstream some 400 km to Chiromo on the Shire. Beyond, the river was unnavigable, and travellers continued on foot up the escarpment to Blantyre, 80 km to the north.

Its name, Shire Highlands Railway, described the line perfectly for to reach Blantyre, the track climbed from an altitude of 40 m at Port Herald (later Nsanje), up 1-in-44 grades over the escarpment to Limbe, at an altitude of 1 250 m, before dropping slightly as it reached Blantyre. Port Herald, to the south of Chiromo, was chosen as the railway's southern terminus as the Shire was slowly silting up and it was assumed that this new 'port' would be safe for years to come. Begun in 1903, the 3′ 6″ (1 067 mm) railway pushed north for 190 km and was opened to traffic in 1908.

The first locomotives were two Bagnall 0-4-0STs, built in 1902 and classed A. Both engines have been preserved, one at Limbe and one at Blantyre. For road service, four small 4-6-0s, class B, were built by Kitson between 1903-07. For the fierce grades, however, eight-coupled power was needed, and a former Rhodesian 7th Class 4-8-0 was bought in 1907. This was followed by the first of the class D 4-8-0s, built by Hunslet in 1910. Sporting a large diamond stack, this engine made no secret of its diet – wood collected in the forests along the right-of-way. In fact, all locomotives on the line relied on this local fuel until 1925 when coal was introduced. In the early days, several old Cape Government Railway locomotives of the 4-4-0 and 2-6-0 types were bought and these elderly engines – some dating from 1875 – probably were used for construction work.

The choice of Port Herald as a terminus proved less favourable than anticipated for at times fluctuations in the Shire's depth left the railway high and dry. To overcome this, a southward extension to Chindio, on the Zambezi, was built. As the new line crossed Portuguese territory, a new company, the Central Africa Railway Co. was formed. Completion of the new 102 km line became a matter of urgency with the outbreak of World War I. At that time Tanganyika was a German colony and part of British strategy was to occupy Lake Nyasa, an objective made easier when the new line was opened in 1915.

After the war, development of the port of Beira in Moçambique made logical a southward extension of the otherwise isolated railway. But the wide Zambezi barred the way and funds for a bridge, or even railway, were not forthcoming from England. However, a group of Belgian financiers decided to take up the projected 250 km railway, running north from Beira to the south bank of the Zambezi. The employees of their Trans Zambezi Railway Co. battled through tsetse fly-infested swamps and forests where bad-tempered buffalo lurked and lions dined occasionally on human flesh. In spite of these natural hazards, the terrain was not difficult and construction, which began in 1920, was completed two years later.

This almost gave Nyasaland a rail link with the sea, but for 13 years there was no uninterrupted through service and paddle-steamers had to link the Trans Zambezi and Central African railways. Then, in 1935, the 4-km Zambezi River Bridge was completed: its approach viaducts and 46 spans of varying lengths making it the longest railway bridge in Africa, and it is recognized as one of the railway wonders of the world.

The fulfilment of this great project led to a change in the managerial organization of the Shire Highlands Railway and the Central African Railway. The former was renamed the Nyasaland Railway and operated both lines. Eventually the Trans Zambezi Railway was taken into the fold as an associate company and all three individual railways were managed by one director based in Nyasaland.

From 1935, locomotives which fitted into this seemingly complex arrangement, though operated by one company, were individually lettered for one of four railways – but numbered in one sequence. Thus, of 14 class D 4-8-0s, bought between 1910 and 1931, two were lettered SHR, two CAR, seven TZR and three Nyasaland Railways.

Three 2-6-2+2-6-2 Garratts, similar to those on the new Cape Central Railway in South Africa were bought for the TZR in 1924 and 1927, but were never very popular and proved this railway's only foray into the world of articulation.

In 1935 the Nyasaland Railway reached the shores of Lake Nyasa then, for the next 35 years, the railway settled into non-expansive lethargy. Traffic increased gradually and more locomotives were obtained – the most interesting being six WD 4-6-2s, built in 1946 to the design of the Sudan's class 220 Pacific – but no new track was laid.

British locomotive salesmen were active after World War II and not all of them were protagonists of diesel. North British capitalized on the original Vulcan 'River' class design for Nigeria, and convinced the Nyasaland Railway that this was the best locomotive for their purposes. No less than 30 of these class G engines were split between the various systems (16 to TZR, 13 to NR, and one to CAR). The first of these arrived in 1948 and the last in 1957, and were both the most numerous single class and the last steam engines purchased. Five years later the first diesels nosed their way onto the rails.

In 1965, soon after Nyasaland became independent Malawi, the railway association was reorganized as Malawi Railways – a combination of NR and CAR with the TZR headquartered in Beira – though one general manager ran the over-all operation. However, this association eventually disintegrated and in 1969 Moçambique's CFM acquired the assets of TZR but continued to operate it as a semi-autonomous system until the change of government in 1975.

Further north, the last day of 1968 saw the first dieselization of Malawi Railways and by July 1969, at Limbe shed, derelict steam locomotives and a Sentinel steam railcar presented a sad picture. However, the shed contained three members of the G class and two others were in use on construction trains over the new eastern extension to the Moçambique border which would link up with CFM's Moçambique division. Technically, the railway was fully dieselized, as steam power was not being used in revenue service. But this did not last long; increasing traffic and diesel failures forced a small return to steam. Three class Gs returned to regular service to be finally withdrawn at the end of 1973. This time dieselization prevailed.

South of the Zambezi, the TZR continued to operate some steam and when last seen, in 1975, several Gs were active at Inhaminga and Dondo Junction near Beira. Many more engines, including the WD 4-6-2s, were retired from service and are lying derelict at Inhaminga.

243. **Malawi's first and smallest locomotives, 'Thistle' and her sister 'Shamrock' were built by Bagnall in 1902. Both have been preserved.**
244. **There is more than a faint resemblance in this Malawian 2-8-2 to a Nigerian 'River' Class engine, for the designs are almost identical. No. 46 was photographed at Limbe, during the steam revival of the early 1970s.**
245. **Central Africa Railway No. 9, built by R. Hawthorn-Leslie was little more than a modernized CGR 7th Class 4-8-0, featuring piston valves, Walschaert valve gear, and superheated boiler.**
246. **D-Class 4-8-0 No. 19 seen shortly before her retirement in 1970. Of 71 steam locomotives operated by the Nyasaland and Trans-Zambesia Railway 55 were purchased new – 14 of which were these 'Ds', second most numerous type on the railway.**

245 246

10 THE SMOKE THAT THUNDERS

BELGIAN CONGO *(now Zaire)*

Neither the railways nor the steam locomotives of this country have ever been adequately recorded and documented. This is partly because the Belgians have never been great railway enthusiasts so that their steam locomotives ran in relative obscurity, and partly because the international 'World basher' fraternity came into being only when much of the world's steam was threatened or being replaced. The then war-torn Congo/Zaire was low on the list of priorities, particularly as the political climate did not encourage the presence of the innocent enthusiast.

Even the exact whereabouts of all the railways is in some doubt. Charles Small has visited and photographed the Kivu Railway, which does not appear in most international directories. Whether this still exists or not is an open question, as is the case with many similar lines. In some countries the railway story is documented down to the last railbolt or locomotive rivet; in the Congo the opposite pertains and the possibility of acquiring good action photographs of Congolese steam trains is probably more remote than a trip to Mars.

Geographically, Zaire is the best watered country in Africa, resulting in an extensive river system, draining westwards from the Ruwenzori mountains into the Atlantic Ocean. These rivers, of which the Congo (now Zaire) is the main artery, fed by the Ubangui, Lualaba, and Kasai, plus many other lesser tributaries, were navigable, except on sections with rapids. Several railways were built merely to bypass these rapids, and even today, all incoming and outgoing traffic requires trans-shipment to and from river boats and barges. Paradoxically, traffic for Zaire landed at ports in Angola, South Africa, or even Maputo in Moçambique, travels by rail, for all these places have direct rail connections to central Zaire.

It was the discovery of copper in Katanga and Northern Rhodesia which accelerated railway building in the territory, established in 1885 and colonized by the Belgians in 1908. Sir Robert Williams was greatly concerned with early railway building in the Congo, and it was largely due to his persistence that what is now the main rail network, the Bas Congo Katanga (BCK), was started. This is one of the few African railways which started from an inland source and reached towards the sea.

Congo railways were neither the result of central planning, nor of inspired forethought – at one time there were eight isolated sections of railway of four different gauges, connected to one another only by rivers.

The main BCK line started at Sakania, on the border of Northern Rhodesia, in 1909, reaching Elisabethville (now Lubumbashi) in 1910, and continuing northwest to reach Bukama, on the Lualaba River, in 1918. The Lualaba is an extension of the Congo, but its unnavigable rapids sections make it almost useless for heavy traffic, thus from 1923 to 1928, the line was extended to Port Francqui (now Ilebo). From there traffic could be sent by river down the Kasai. In 1931, the connecting link from Tenke to Dilolo, on the Angolan border, provided the most direct possible outlet, via Lobito Bay and the Benguela railway. This 3′ 6″ (1 067 mm) system, had a branch to Kabongo.

Meanwhile, the lower rapids of the Congo were bypassed by the 750 mm-gauge Chemin de Fer du Congo, from the seaport of Matadi to Leopoldville, now Kinshasa. Until the Benguela link was completed, this little railway handled most of the country's imports and exports, and eventually was converted to 3′ 6″ gauge. Across the river from Matadi was the Vicinaux, or local railway, of the Mayumbe region. This ran from Boma on the river, to Tshela, but appears to have been closed.

In its semicircular sweep through the country, the Congo/Lualaba River is torn by further sets of rapids, and these were bypassed by the lines of the delightfully named Chemins de Fer du Congo Superieur aux Grands Lacs Africains (Upper Congo-Great African Lakes Railways), a fairly accurate title, as the line traversed the upper reaches of the Congo and reached Lake Tanganyika. Although it could be expected that this railway would eventually connect with the 3′ 6″ network, it was built to metre gauge. Working upstream, the first section was from Stanleyville to Ponthierville (now Kisangani to Ubundu) which remains isolated and metre gauge.

The next section was from Kindu to Kongolo, and after a short river section a line ran eastwards from Kabalo to the shores of Lake Tanganyika at Albertville (now Kalemie). The Kongolo-Kabalo gap was soon closed and in 1955, when Kabalo was joined to the Kabongo branch of the BCK, the gauge was altered to 3′ 6″. Another isolated 3′ 6″ line, built about 1930, was the Chemins de Fer du Kivu, intended to connect Kalandu, on the shores of Lake Tanganyika, with Bakavu, on Lake Kivu. However, difficult terrain prevented the line from proceeding beyond Kamanyola, and it is not known whether it still exists.

The final railway is the Chemins de Fer Vicinaux de l'Uele, an extensive 600 mm system running from Uele River at Bondo, to several destinations, and having a total route length of 1 235 km, probably the world's longest railway of such narrow gauge.

Industrial Railways

The major industrial user in the Congo was the Union Miniére du Haut Katanga which at one time had more than 50 locomotives spread between its plants at Jadotville, Elisabethville, Etoile du Congo and other locations. A catalogue, issued in 1934 by the locomotive building firm of La Meuse, lists all their deliveries round the world, and for the Congo are listed locomotives for oil wells, coal mines, docks, sugar plantations and other industries. One entry is for 14 locomotives for Compagnie Belge de Chemins de Fer et d'Entreprise, which may have been a private railway, or possibly a dealer and exporter. The three other users with more than one or two locomotives were: Charbonnages de la Luéna (coal mines), with five locomotives; Compagnie Sucriére Congolaise (sugar-cane), with seven locomotives; and Société Manutention dans les Ports du Congo (dock company), with five locomotives.

Chemins de Fer Bas Congo au Katanga (BCK)

This railway was built in reverse order of its title, for it was commenced in Katanga, and has yet to reach the lower Congo. It was owned by a holding company, named in the correct sequence, the Chemin de Fer Katanga-Dilolo-Leopoldville, which outlined the route of the railway, although the section from Ilebo to Kinshasa remains incomplete. A total of 166 steam locomotives appear to have been supplied to the BCK, and the 1959-60 edition of the *Railway Directory and Year Book* shows 164 locomotives, though dieselization had started, and probably none remains in service today. Belgian builders provided almost all the locomotives used, half of which were rather uninspiring 2-8-2s. Only 43 years separated the first and last new steam locomotives for the BCK, and at the time of delivery of the final steam engines, all the previous purchases were still in stock. The earliest were wood burners, but the later and larger locomotives burned coal, probably from local mines, though possibly from Rhodesia.

The first five engines, which undoubtedly assisted in the construction of the line, were small 2-6-0Ts by Tubize, built between 1910-13, and were probably used later on light shunting duties. They were followed by 18 larger 2-6-2Ts, from Tubize, La Meuse, and Haine St. Pierre, built between 1912-25, and the final tank engines were 11 0-6-0Ts for purely shunting duties, again by assorted Belgian builders, built in 1922-28. Except for the length and number of wheels, all these tank classes were rather similar in appearance, with diamond-type, spark-arresting chimneys, and wood fuel racks on top of full length side tanks.

The first proper main line locomotives followed Southern African standards and were 4-8-0 tender engines; though generally modern in design, with piston valves and wide fireboxes, they were unsuperheated, although

247

superheaters were later applied. Twelve came from Tubize in 1914, and seven more were built in 1924, some by Cail. During World War I, a batch of 21 2-8-2s were supplied by the American firm of H.K. Porter. These were shipped via South Africa, which was itself short of power, and borrowed four for a while, using them on SAR metals. These also were not superheated but, as with the 4-8-0s, superheaters were later fitted.

The Porter engines evidently persuaded the Congo authorities that the 2-8-2 was a more suitable design than the 4-8-0, for until immediately prior to World War II, this was the predominant type supplied. Sixty such engines, of three types, were built by a wide range of Belgian builders, though it hardly seems economic for the 16 101 class engines of 1924-29 to have been built by four firms, each having to tool up for a few engines.

Except that they had piston valves and superheaters, these 101 class were almost identical to the Porter engines, and their construction was shared by Cockerill, La Meuse, Tubize and St. Léonard. The 301 class again was similar, but had plate instead of bar frames, and 22 were built in 1921-26 by Franco-Belge, Haine St. Pierre, and La Meuse. The final version came out in 1929-38, and reverted to bar frames, but with a longer wheelbase. These were the 401 class, again 22 locomotives, by Haine St. Pierre and Franco-Belge.

Traffic seems to have expanded rapidly on the BCK, for the locomotive stock increased both in quantity and in power. In 1938, Cockerill built an experimental, but rather clumsy 2-10-4, no. 801, which was not repeated. In 1939, La Meuse built two rather more respectable looking 4-8-2 mixed traffic engines, which were apparently successful, as 16 more were supplied in 1947. Also in 1947, Cockerill built another ten-coupled engine, a 2-10-2, utilizing the same boiler design as the 4-8-2, but again this remained a solitary engine. Possibly, at that time, the track was not good enough for a ten-coupled design, although with very small wheels, the coupled wheelbase was far from excessive.

Eventually, BCK followed the rest of Africa and ordered Garratts, 12 being built in 1953 to the same design as those supplied to Moçambique the following year, both batches being built by Haine St. Pierre. Exactly where the various locomotives worked, especially the larger types, is not known with certainty.

247. **Looking down on the wharfs, trans-shipment sheds, and locomotive depot at Matadi, gateway to the Congo and Central Africa. This photograph was taken after the line was rebuilt to 1 067 mm gauge, and shows a group of 0-6-0T and 2-8-2 tender engines awaiting duties.**

248. **The Mayumbe Railway, on the north bank of the Congo River, introduced the Garratt to Africa with this diminutive 0-4-0+0-4-0, No. 1A, shown at the maker's works.**

248

249

Chemins de Fer du Congo Superieur aux Grands Lacs Africains

Though the precise opening dates of this railway's three sections have not been ascertained, that from Stanleyville to Ponthierville was the first, in 1908, and construction must have begun about 1903, when the first locomotive was supplied. The earliest locomotives were all tank engines, starting with some 0-6-0Ts by La Meuse in 1903, obviously used for construction work. Tubize built three 0-4-0Ts in 1904, and the following year supplied two 2-6-0Ts, which would have been suitable to start services in 1908. One assumes that as one section was completed, the construction locomotives moved on to the next – with the exception perhaps, of two or three left behind for shunting and similar duties. Between 1906 and 1911, 13 more 0-4-0Ts were bought from Tubize, St. Léonard, and Couillet, as well as two 0-6-0Ts – presumably mainly for construction work, being unsuitable for the main line.

The first main-line engines came from Tubize, who built 20 small 2-6-0s with four-wheel tenders, between 1913 and 1924, as well as four more 0-6-0Ts. Traffic must have been light to be handled by these little Moguls, but must have increased fairly rapidly, for Haine St. Pierre then built six 2-8-2s, a pair in 1926 followed by four in 1930, and even a pair of 2-10-2s in 1937 – the largest steam locomotives to run on the system. No more new engines were obtainable during the early part of World War II, but two Porter-built standard American army type 2-8-2s arrived in 1944. The final new steam was six 2-8-2s by Haine St. Pierre in 1950, though subsequently a few of KDL's 101 class 2-8-2s were also transferred to the Grand Lacs system, when displaced by dieselization. The Grand Lacs lines probably still retain some active steam; some steamers were noted by a visitor in 1975, and Zaire still lists a few in their annual returns, though not detailing their whereabouts.

Chemins de Fer du Congo

This remarkable railway was – and had to be – the first to serve the Congo and its tributaries, for it bypassed the lowest series of rapids on the river and without it no worthwhile traffic could reach the navigable sections beyond the Stanley Pools. Starting at Matadi, on the lowest section of the Congo accessible by ocean-going vessels, the line was as far seawards as possible while remaining in Belgian territory. The final stretch of river formed the boundary with Angola, while at Stanley Pools, the north bank was in French Equatorial Africa. The 750-mm gauge chosen was presumably thought adequate in 1889, when construction started, but was later found to be a short-sighted decision.

Although nominally of 750-mm gauge, to which rolling stock was built, track was laid to 763-mm because the native labour available for construction was considered too primitive to understand the principles of gauge widening on curves. Pre-punched steel sleepers were used, with gauge widened throughout. As most of the track was on curves, ill-effects will have been minimal, but the first little locomotives must have yawed alarmingly when they reached a stretch of straight.

All the earliest locomotives were tank engines from St. Léonard, and there seems to have been no attempt at standardization – possibly they were stock types acquired as available. The first three were 0-6-2Ts, dated 1890, of which two more came out in 1893-94. Between 1896 and 1902, 12 0-4-0Ts were supplied, while 24 0-6-0Ts arrived between 1894 and 1897, to total 41 locomotives. The next batch of engines, also 0-6-0Ts were numbered 31 to 40, so possibly some of the earlier engines, particularly the 0-4-0Ts, were regarded exclusively as construction machines. This collection of tank engines sufficed for a decade, but from 1910 traffic grew rapidly. From 13 000 tonnes handled in 1898-99, it had risen to 79 000 tonnes in 1913-14, and to cope with this, 27 more 0-6-0Ts were built in 1910-11.

Clearly, there was little point in flooding the line with more miniscule engines of low haulage capacity. On the coastal end, particularly where the escarpment was ascended fairly rapidly, a single-track railway of such narrow gauge and light rail would need several locomotives per train to handle a reasonable amount of traffic, and it is doubtful whether the necessary number of engine crews would have been available.

Substantially greater motive power was needed, and this pointed to articulated power. The normal continental answer to this problem, at the time, would have been the Mallet, but the solitary Mallet on the Belgian State Railways, for banking at Liége, was not considered particularly successful. At

250

249. **The C.F. Bas Congo-Katanga, following earlier African railway practice, had a batch of 4-8-0 tender engines. No. 1011 is one of these British-looking, Belgian-built locomotives.**
250. **The 2-8-2 type was introduced to the Congo with a batch of American locomotives built by H.K. Porter. Proving satisfactory, further engines of this type were produced in Belgium, typified by No. 413.**
251. **Katanga Railway's experimental 2-10-2, No. 901, a heavy freight hauler which never passed prototype stage. This locomotive incorporated the 4-8-2 boiler design married with the earlier 2-10-4's running gear.**
252. **Another heavy freight prototype on the Katanga line, huge and clumsy looking 2-10-4, No. 801, on test at her builder's factory.**
253. **For the heaviest duties on the Matadi-Leopoldville line, Otraco followed Belgian main-line practice by ordering the very European, and equally non-African, 2-10-0 type, as No. 254 illustrates.**

251

252

253

254

255

256

254. C.F.A. Kivu, operated by Otraco in the far eastern Congo, used this typically Belgian wood-burning 0-6-0T for shunting on the wharf at Kalandu in 1955.
255. Main-line power on the C.F.A. Kivu comprised two dinky wood-burning 2-8-2s, one of which is shown at Kalandu in 1955.
256. The only successful large engine design in Katanga was the 700 series 4-8-2. No. 706 is shown in ex-works condition at Mutshatsha in 1972, after its final overhaul.
257. A month before the gauge was widened from one metre to 1 067 mm, a 0-6-0T with auxiliary tender shunts at Albertville on the Chemins de Fer Du Congo Supérior aux Grands Lacs Africains, in August 1955. This is believed to be the last railway in Zaire to operate steam locomotives, and in July 1973 this type was one of several still at work.

that time, Beyer Peacock were introducing the Garratt; St. Léonard, realizing the potential of this double-jointed steam design, took up manufacturing rights under licence. They almost came unstuck on the first engine for the C.F. du Congo, in 1911, which included an unconventional boiler and gave trouble. Fortunately, the other good points of the Garratt design were recognized, and when a normal boiler was applied, the now useful unit was able to haul double the load of the previous 0-6-0Ts. As a result, a further 31 Garratts were built in 1919-25, forming one of the Garratt's great early successes. At the same time, 23 larger 2-6-2T tank engines were built by Cockerill.

In the absence of any detailed operating data on this early railway, in an unhealthy tropical climate where one in five of the 4 500 construction workers succumbed to malaria and other diseases, we can only assume that the Garratts were used on the coastal section where the Palabala escarpment meant a climb to 272 m altitude in 16 km, after which the 2-6-2Ts probably managed the same load inland to Leopoldville where river transport was resumed. The final locomotives for the narrow gauge were five Tubize 2-8-2 tender engines, with outside frames, built in 1926.

By then, the limitations of the 750-mm gauge could be ignored no longer; and there was the possibility of linking the railway to the Cape gauge line from Katanga coastwards. In 1921 it had been decided to widen the line to Cape gauge – easier gradients and curves would give an annual capacity of 1 300 000 tonnes, compared with the 280 000 tonnes of the narrow gauge, and train loads would be trebled.

The result was a completely new railway, generally along the same alignment as the narrow gauge – though some distance apart in places – and crossing it six or seven times. Work began in 1923 and was completed, in three sections, by 1932. During construction, both 500 mm- and 750 mm-gauge railways were laid down, the former using 16 locomotives – probably those used to build the narrow-gauge line three decades earlier. How the change from narrow to Cape gauge was effected is not clear – was the broader gauge brought into use in sections, with consequent clumsy and troublesome transshipment, or were the two lines operated side by side until the broader gauge could handle all the traffic? The gradual acquisition of Cape-gauge motive power seems to suggest the latter, but it is unclear when the narrow gauge ceased to function. Some of its later motive power may have been converted to the broader gauge, though there is no evidence of this.

The locomotives built for the broader gauge were of two main types – heavy power for the escarpment section, and lighter engines for the highland. In 1930, Tubize built five light 2-8-2s and St. Léonard built three heavy 2-10-0s. The following year six light engines from Cockerill and five from Haine St. Pierre were added to the roster, the latter company supplying a final two in 1951. Meanwhile, the heavy 2-10-0s had increased slowly with three from Franco-Belge in 1938, and a final six from Haine St. Pierre in 1947. By 1960, the *Railway Year Book* recorded a total of 60 steam and 51 diesel locomotives, which would indicate some conversions from 750 mm to 1 067 mm gauge. By 1965 the line was reported to be entirely dieselized, and whether sufficient records survived the Congo unrest to establish the full story of steam days remains an enigma. Better records of the 1860s exist for some parts of the world than those for the Belgian Congo 20 years ago.

Chemins de Fer Vicineaux du Mayumbe

This little railway on the north bank of the Congo estuary, was the westernmost railway in the Belgian Congo. In 1898 a private company obtained a 99-year concession to build a narrow gauge railway through the fertile lands extending towards the Mayumbe mountain range in French Equatorial Africa. Instead of the more probable 600 mm gauge, the line was built to 610 mm. Occasionally it is referred to as being 615 mm gauge but possibly this, like that of the C.F. du Congo, refers to the widened gauge on curves.

The line ran almost due north from the river port of Boma, and soon faced financial difficulties. By the end of 1901 it had penetrated 81 km to Lukala, and the state agreed to a temporary halt; in 1910 a further 56 km to Tchela was begun and this, opened in 1912-13, remains the terminus, though from time to time further development has been proposed.

257

The railway was unusual in that all its locomotives were from one builder, St. Léonard of Liége, which possibly had a financial interest in the line. The first eight engines were minute 0-4-0Ts, built in 1898-99, which sufficed for the first decade of operation. The limited loads which these few small engines could have handled, must account in some way for the company's financial difficulties, in spite of the primitive alternative forms of transport.

For the increased traffic expected from the Tchela extension, St. Léonard built six small 0-4-0+0-4-0 Garratts in 1910-11. Though the Garratt was a British invention, developed by a British firm when much of the railway development in Africa was in British colonies, these were the first in Africa – the continent where the Garratt really made its home. This minor Belgian colonial railway introduced the product to its largest market, and all subsequent locomotives built for the line appear to have been Garratts, although one locomotive is indeterminate.

The original six were Class A, and these were followed by six class B in 1919-21, and five more class B in 1924, making this the most prolific series on the C.F. Mayumbe. Four class Cs were built in 1926, and a solitary class E in 1927, all being 0-4-0+0-4-0s, similar in size but differing in detail from the class A.

Class D was a mysterious 0-6-6-0T built slightly earlier than the class E (works number 2093 compared with 2096, both dated 1927). However, the St. Léonard Garratt catalogue which illustrates the other classes omits the D. Was it a six-coupled Garratt, which seems likely, or some other form of articulated steam locomotive? Until an illustration of Mayumbe's class D is unearthed, this will remain an intriguing mystery.

Chemins de Fer Vicineaux de l'Uele (later Vicicongo)

This was one of the world's longest railways of so small a gauge, perhaps second only to the 600 mm-gauge system in Morocco which it outlived by many years. Its name abbreviated to Vicicongo, this railway was a latecomer to the Congo railway network, authorized only at the end of 1923. As so often in the Congo, it was to link navigable sections of the river network – in this case to provide a connection between the Congo River and the Uele River, on a stretch of which rapids prevented suitable river transport. As the Itimbiri River was navigable from its confluence with the Congo as far as Aketi, an inter-river light railway was constructed though the jungle to Bondo on the Uele River. As a local link for light traffic its eventual ramifications were probably unforseen. From Jamba, a branch led off to Titule, and from Lienart on this branch, a considerable extension was built to the terminus at Mungbere.

In the early 1920s a large quantity of light equipment was available from the former military trench railways, and Wiener records that in 1928 the line had 20 12-tonne locomotives with Klien-Lindner axles. The handful of extant photographs from this era show that these were ex-German Feldbahn locomotives, though they provide no details of builders or dates.

Tubize supplied the new motive power, and in 1929 built six 0-6-2Ts of new design with wide firebox behind the coupled wheels, suitable for burning the local wood.

Though they were built in 1929 it should not be assumed that the engines went into immediate service. They would have been assembled, steamed, and tested before being dismantled for the complex journey, in their crated sub-assemblies, to Matadi. From there they would have been railed over the congested Chemins de Fer du Congo to Stanley Pools, where they would have been transferred to boats or barges, for the next stage up the Congo River. Thus locomotives built in 1929 would not have entered service before 1930 or even 1931.

A further eight 0-6-2Ts were built by Tubize in 1933, and the same firm, in 1937, turned out a much larger design, a 2-8-2 with side tanks and tender. All fuel was carried on the tender, so the tank-top wood racks featured on the 0-6-2Ts were absent. Because of the length of the boiler, a combustion chamber was incorporated in the new design, and this must have helped in consuming the wood fuel, although diamond-type, spark-arresting chimneys were again fitted.

When, in World War II, it was necessary to ferry military supplies to the Eighth Army in North Africa, a river/rail link via the Congo was mooted, as the Mediterranean was unsafe. It was planned to extend the Vicicongo line down to Bumba, on the main Congo River, and to build a considerable extension at the eastern end to join up with the Sudan railways. To cope with the expected extra traffic, 18 of the line's 2-8-2T + tender engines were ordered from W.G. Bagnall, but Montgomery's victory over Rommel at El Alamein made the trans-Africa rail scheme redundant and only five were completed and sent to the Vicicongo.

Some completed components were taken to a War Department supply dump near Derby, and some of these eventually may have found their way to the railway as spares.

RHODESIA *(now Zimbabwe)*

Landlocked Rhodesia (now Zimbabwe) was, and is, completely dependent upon rail outlets to the ocean through neighbouring countries, and work on the first two such lines commenced almost simultaneously. The shorter but more difficult route was to Beira on the Indian Ocean, and was from the Pungwe River, up the Amatongas escarpment to Umtali – a 2-foot (610 mm) gauge line commenced in 1892. Difficulties were many, and the gauge chosen so inadequate that the narrow gauge to Umtali was not opened until 1898. It was converted to Cape gauge in 1900.

Meanwhile, the longer but easier Bechuanaland Railway was commenced from Vryburg, in the Cape of Good Hope, through Mafeking and Bechuanaland to Bulawayo, which was reached in 1897. Relatively easy terrain allowed construction to proceed rapidly and within ten years a Cape gauge artery existed from Vryburg to Beira. The main north line reached Victoria Falls via Wankie in 1904; the Zambesi was bridged in 1905, and Broken Hill (now Kabwe) attained in 1906. Ndola, centre for the Copperbelt and on the Congo border was connected by rail in 1906, less than 20 years from the beginning of a Rhodesian railway system.

The third outlet to the sea, via the Benguela Railway in Angola, took 23 years (from 1905 to 1928) to reach the Congolese border, from which it traversed Katanga to connect with the Rhodesian system near Ndola. By this time, Rhodesia had outlets to the Indian Ocean, via Beira in Moçambique, to other Indian Ocean ports, such as Durban in South Africa, and to the Atlantic at either Cape Town or Lobito Bay – with the two most direct lines in Portuguese territories.

Locomotives

The narrow gauge locomotives built for the Beira-Umtali line were really temporary – the line soon being widened – and ran almost entirely in Moçambique. The majority were 4-4-0 tender engines, built by Falcon of Loughborough, and when the railway was re-gauged after two years, most of these engines were used elsewhere in southern Africa, and several exist today.

258 (Previous page). **Running 90 minutes ahead of time, sufficiently early to catch the chill of a Rhodesian winter's morning, 14A and 16A Garratts combine to heave a load of limestone through Mulungwane Gorge on the West Nicholson branch in July 1975.**

259. **The Victoria Falls bridge was built with two lines of railway. During reconstruction, when one rail line was replaced by a motor road, an 11th Class 4-8-2 on the mail train posed on the bridge for a magnificent panoramic portrait.**

The earliest 1 067 mm-gauge locomotives were mainly of the Cape Government type. The class 6 4-6-0 was ordered for the Bechuanaland Railway but actually delivered to the CGR, and, as far as is known, did not run in the Rhodesias. There were also two 4-4-0 locomotives, built in 1891 and acquired by the Mashonaland Railway in 1897, but these were too small for general use. Two other early oddments were a pair of 4-6-2Ts built for the Sea Point line in the Cape, and sold to the Mashonaland Railway in 1897. The subsequent name 'Rhodesia' did not appear on these early locomotives, the line from Beira becoming the Mashonaland Railway, and that from Vryburg being the Bechuanaland Railway. The Bechuanaland Railway became Rhodesia Railways in 1899, by which time it was in Matabeleland, and in 1901, the clumsy title of 'Beira and Mashonaland and Rhodesia Railways' was introduced. The title Rhodesia Railways came into official use in 1927.

The most numerous of the early engines were the Cape 7th class, a light and flexible 4-8-0 well suited to conditions on pioneer railways. Nearly 70 of these were placed in service between 1897 and 1903, some of the later examples having Belpaire boilers. Between 1914 and 1923, 13 of these were rebuilt as 4-8-2Ts and 4-8-4Ts for shunting duties and were classed 6 and 6A respectively. It is noteworthy that the RR classification derived directly from that of the CGR, commencing basically with class 7. Apart from the rebuilds to tank engines, none of the lower class designations were taken up – though various engines could have been classed 1 to 5, but in the event remained unclassified. One of the old 7th class locomotives, No. 43, remains in working order at Bulawayo Railway Museum, and is occasionally steamed for special occasions. A rebuilt tank engine is likewise preserved, but as a static exhibit. These two engines date to the first decade of railway operation in Rhodesia.

Another Cape design, this time the 8th, was chosen when increased power became necessary, and 17 of these well-tried 4-8-0s were built from 1904 to 1910. These were a reversion to straight, unarticulated machinery following the trial purchase of two 0-6-6-0 Kitson-Meyer engines in 1903 – identical to two others in South Africa, for the Cape and Central South African railways.

Basically, the Kitson-Meyer was a logical type of articulated locomotive, which, like the later Garratt, had plenty of space for a free and uncluttered firebox and ashpan layout, but the design evolved for Africa found no favour on any of the three railways which tried it, and all were scrapped after a short life. The main weakness in these early Kitson-Meyers was that they exhausted the steam from the rear cylinders directly into the atmosphere, thus providing no help in inducing draught through the usual smokebox arrangement. This led to poor steaming and heavy coal consumption, and was only rectified in later examples for South America, where the type achieved fair popularity, being built until 1935 for Columbia and existing in service at least until 1977 in Chile. Fortunately, experience with the Kitson-Meyers did not totally discourage Rhodesia from trying articulated power, and in due course the Garratt proved completely successful.

The first class which can really be called a Rhodesian design was the 9th class, basically an enlargement of the 8th class, but with superheater, piston

valves and Walschearts valve gear, plus a wide firebox, to give an altogether more modern design. Eighteen were built in 1912, followed by six in 1915, and the type was soon adopted by the Benguella Railway in Angola, which bought 30 of the same design from 1913 to 1930. Both railways also bought locomotives from America during and after World War I, but these differed – Rhodesia's class 9A coming from Alco in 1917, while Benguella bought their 9B class from Baldwin in 1921. The latter should not be confused with RR's own 9B class, which were large-boilered rebuilds from the original 9th class. Twenty-six of these rebuilds were carried out in Bulawayo works from 1939 onwards, almost making the original 9th class extinct. The 9Bs were chunky and handy machines, and survived in branch-line service until the mid-1950s, especially from Gwelo to Shabani. Later they were relegated to shunting, finally being withdrawn in the 1970s after 60 years' hard work.

Mountain classes

The name 'Mountain' originated in the USA to denote locomotives of the 4-8-2 type, the first American examples of which were built for mountain passenger work in 1910, though the 4-8-2 tender engine originated in Natal on 1 067 mm gauge with six conversions from 4-8-0s, in 1906 and the first new 4-8-2s were built for New Zealand in 1907.

In Rhodesia the type offered greater boiler capacity than the previous 4-8-0s, and until large numbers of Garratts became available, was very much the backbone of the fleet.

The first Mountains were the 10th class for mixed traffic duties, and these were clearly based on the CGR design – unsuperheated, and with slide valves driven by inside Stephenson gear. In 1913, North British produced an updated version with superheater, piston valves, and Walschaerts gear, and these became the SAR class 4A and RR 10th class. There were several differences between the RR and SAR batches, notably that the SAR locos retained the raised firebox of the earlier class 4, and that the valve gear was arranged with return crank leading – instead of the more usual trailing position adopted by RR. The RR locos were the only engines on the system with combustion chambers, and the 20 built in small batches from 1913 to 1930 were long associated with the line through Bechuanaland.

The next 4-8-2 to appear was a slightly heavier and smaller-wheeled design, suitable for the heavier sections of the line. Built by the Montreal Locomotive Works in Canada during post-war locomotive shortage, they were contemporary with and similar to, the class 14C built for South Africa, although again, there were several important detail differences between the RR and SAR versions. The 30 RR 11th class were used extensively in the Copperbelt of what was then Northern Rhodesia. The last survivors on RR metals were used for heavy shunting, but six were sold to Moçambique in 1962, and may still be in use. After World War II, the locomotive shortage again led to 12 similar engines, to this by then outdated design, being built by Montreal in 1948. These became class 11A and, after a fairly short life in Rhodesia, were all sold to Moçambique in 1962.

Rhodesia's most popular 4-8-2, both in terms of quantity and in enginemen's esteem, was the 12th class, of which 51 were built by North British between 1926 and 1930. This was an intermediate design, smaller than the 11th class, and more generally useful than the 10th, and very much a general purpose engine, equally at home on freight or passenger work, main line or branch. Of the first 20, ten had oscillating-cam poppet valves, operated by Walschaerts gear, but this was not beneficial, and all reverted to piston valves standard with the other ten. Until 1978, two or three 12th class were used on shunting at Thomson Junction and Wankie. When en route to and from Bulawayo, once a fortnight, these hauled main-line freights. Now restricted to shunting around Bulawayo, they are seldom seen on the main line.

In 1943-44 three of the 12th class were rebuilt at Bulawayo, with larger 11th class boilers, becoming class 12A. Boilers were built for ten more conversions, but were used instead for ten new engines, class 12B, built at Bulawayo in 1953 – the only locomotives built in Rhodesia. The rebuilds and new engines worked most of their RR lives on the Copperbelt, and, apart from one 12A destroyed in a collision, were all sold to Moçambique in 1962, where, with the 11A class, they were particularly associated with the Swaziland Railway.

Rhodesia's final Mountains were all of SAR design. Twenty 19th class, built by Henschel in 1951, were the first Rhodesian locomotives not supplied from a British or Commonwealth source. These were virtually identical to the SAR class 19D, the main discernible difference being in the tender bogies, which had plate frames instead of the cast 'Buckeye' type. Unlike South Africa, where the 19D is popular with enginemen, the RR 19th class has never been liked, generally being considered inferior to the older 12th class. An additional engine of the same type, but with condensing gear, was supplied by Henschel in 1954, and classed 19C. Not surprisingly, trouble was experienced with a 'one-off' job, but when allocated to a regular crew, satisfactory work was obtained on the long dry run to Malvernia, on the Moçambique border. After a collision in 1958, the engine was rebuilt as a standard 19th class and continued in service until recently.

Two more basically similar engines came from the Nkana copper mine in 1968. These had been built by Henschel in 1952 and were surplus when the mine dieselized. The spread of dieselization on RR again made them surplus and, in 1974, they were sold to Selebi Pikwe mine in Botswana, where they are to be joined by a further two of the original batch.

The final 4-8-2s to run on RR were all ex-SAR, and comprised firstly six class 15E heavy locomotives with poppet valves, which were hired from South Africa and later purchased by RR. As dieselization proceeded, these were scrapped. Then, in 1978-79, six of the similar 15F were hired. For some reason, these, the most popular engine in South Africa, failed to find favour with RR crews, and were returned after less than a year's service.

260. **Easter 1976, and a 12th Class 4-8-2, piloting a 15th Class Garratt, rolls into New Wankie with a southbound coal train from Thomson Junction.**
261. **Long after they had ceased work in Rhodesia, the 16th Class 2-8-2+2-8-2 Garratts could be seen hauling trains on the Benguela main line in Angola, and on South African collieries. Maroon-liveried 618 is seen hauling empties upgrade at Enyati, Natal.**
262. **One of Rhodesia's splendid 15th Class 4-6-4+4-6-4 Garratts takes water at dawn with a main-line mixed train.**
263. **In 1975, before replacement by a Garratt, 12th Class 4-8-2, No. 177 ambles round the back of Wankie drawing a light load to an industrial siding.**
264. **15th and 20th Class engines, rebuilt by RESCCO under Zimbabwe's current rehabilitation scheme, are being given attractive names of local Matabele origin. Cab-side detail of No. 419 'Isambane', 15A Class.**
265. **Intertwined 'RR' initials are to be found on smokebox sides of most Rhodesian locomotives, following the 'MR' example of the former Mashonaland Railways.**
266. **The driver of 20A No. 727 peers out of his cab.**

		262
260	264	263
261	265	266

ISAMBANE
419
15

RR
W B
BOILER Nº 801

727
20A

Land of Garratts

Rhodesia is second only to South Africa in the total number of Garratts supplied to the system – with 250 RR Garratts by comparison with about 400 in SA. Whereas South Africa's Garratt fleet never amounted to a sixth of its total locomotive stock, nearly half the locomotives built for Rhodesia have been Garratts, which for many years have been predominant. And, whereas RR's straight locomotives have often been of, or closely derived from, South African types, RR Garratts have, with two exceptions, been exclusively designed for Rhodesia.

Following the failure of the Kitson Meyers, RR's first Garratts, the 13th class, were a bold step in that 12 engines were ordered simultaneously, instead of one or two for experiment. In 1925, South African experience – often followed by Rhodesia – was limited to a single main-line Garratt and several branch and narrow-gauge engines, so that the only really comparative SAR class – the GD – was not in service when RR's 13th class was ordered. In some ways it makes sense to order a dozen new locomotives, rather than one or two, for a dozen provide for greater crew familiarity and, with it, more chance of success. And, whereas it may be possible to write off one or two engines which seem not to succeed, a dozen engines *have* to be *made* to work. Whether or not this was the thinking at the time, these first 13th class proved successful, though not without aspects in need of development. Plate frames were a weak point in the design, and two engines with poppet valves were later given piston valves to conform with the rest. The Garratts were used on the most severe section of what was, for a short time, still the Mashonaland Railway, from Vila Machado in Moçambique, to Umtali near the border.

A further 16 Garratts were supplied, again by Beyer Peacock, in 1929-30. Though of generally similar dimensions they had bar frames to provide a more robust chassis; the boiler centre was raised, and a sloping grate incorporated, which together gave better combustion and more ashpan capacity. The original Beyer Peacock square front water 'cistern' and low bunker were replaced by a tank with rounded top corners, and deeper, self-trimming bunker, giving the appearance of a larger engine. Although now 50 years old, the 14th class has survived well, the eight earliest engines being sold to Moçambique in 1949 where, nominally at least, they remain in service at Gondola. One of the second batch is in Bulawayo museum, and of the remainder laid aside in the early 1970s, three have recently been returned to service for shunting at Bulawayo.

The next Garratt class did not take up the next vacant number, but became the 16th class. A direct development of the earlier locomotives, which were of the 2-6-2+2-6-2 type, the 16th class was a 2-8-2+2-8-2 having 33% more power. These were ordered for the heavy coal traffic from Wankie to Livingstone, and Beyer Peacock built eight in 1930, and a further eight in 1938. After realignment and grade easing of the Livingstone line, they were largely reallocated to the Umtali section, and, when made redundant in the 1960s, nine were sold to the Benguela Railway service, the only survivor in Rhodesia being the museum example.

The vacant 15th class designation was filled in 1940 when Rhodesia's most useful and numerous locomotive class appeared. It was a 4-6-4+4-6-4, a type used previously only in Sudan, and used the same boiler as the 16th class, becoming, therefore, a general purpose version of the previous freight slogger. Good cylinder and valve design, coupled with the largest driving wheels in Rhodesia, made it a speedy machine, easily able to reach 110 km/h (70 mph) when allowed to run. A total of 74 of these excellent machines was built from 1940 to 1952, all by Beyer Peacock, except the last ten which were built in Belgium under licence. As may be expected in so numerous a class, several variations appeared, of which the most obvious was the front tank shape. The initial four locomotives had a very rounded design with no square edges, and these were followed by a batch of ten with tall, wide tanks, but with the front section partitioned off and the same capacity. The leading two edges of these tanks were square-cornered. The remaining 60 engines had lower tanks of basically the same features as the second batch, and this is now 'standard', some of the earlier engines being fitted to this design when new tanks were needed. The bunker shape of the three batches also differed.

The final 40 engines arrived with boilers of higher pressure, and were classed 15A, but as these have been interchanged throughout the class, one cannot now tell a 15th from a 15A by engine number. Two boilers were fitted with Giesl ejector front ends, of which one also had a superheat booster. These circulated onto other locomotives, and eventually there were four Giesl 15s, although only two at any time. The Giesl ejector gave good results when properly aligned, but tended to distort out of alignment in service. This, coupled with the imposition with sanctions on Rhodesia, led to the experiment being dropped and substituted by a locally designed multiple jet chimney which gave consistently good results – not as good as a 'good Giesl', but better than a 'bad Giesl'. One 15A, No. 406, was fitted with two thermic syphons, which it retains today, although the engine is out of use.

The 15/15A series was used extensively throughout the RR, though mainly based in Southern Rhodesia, and run to Mafeking, Salisbury, and Livingstone/Victoria Falls. One engine had a tragic history with an amusing end. No. 404 was involved in several mishaps and derailments, eventually killing

its driver in a derailment at Lukosi, near Wankie. It was then considered a 'hoodoo' engine, and drivers were reluctant to take it out until someone exorcized it by renumbering it 424, since when it has caused no problems. The bent numberplate remains bolted to a sleeper at the site of 404's final accident, and '404 curve' is known among keen photographers as a fine location for shooting Garratts.

After the introduction of the 15th class, heavy wartime traffic created an urgent need for more Garratts and nine of the heavy type WD Garratts were allocated to RR, becoming the 18th class. As a stop-gap they were a useful acquisition, being even more powerful than the 16th class, which they assisted with the Wankie coal traffic; but, being plate-framed engines, maintenance was heavy, and they were sold to Moçambique where they hauled goods and passenger trains between Gondola and the border.

The post-war decade of 1948-58 saw RR commission 109 Garratts of three new classes, all of modern design, in addition to the 70 post-war examples of the 15th/15A class. All were built by Beyer Peacock. For branch-line service, there were 18 of class 14A, built as a modernized version of the 14th class, with whose basic dimensions they were identical. This showed that the dimensions chosen for the original 13th class of 1925 were a happy selection, still being suitable for branch-line engines 20 years later. The 14A class was used almost entirely in Southern Rhodesia, particularly on the branches from Gwelo to Fort Victoria and Shabani, and from Bulawayo to West Nicholson. In the general rundown of steam in the early 1970s, they tended to be relegated to shunting, but are now being refurbished.

To supplement the 16th class, a similar redesign was carried out, producing the 16A class of which 30 were built in 1953, the same year as the 14As. The 16As generally took over from the 16th class and the 18th class, as these were disposed of, and for most of their lives have worked such duties as the branches from Salisbury, until dieselization, and the Bulawayo-Colleen Bawn section of the West Nicholson branch. Eleven were in Northern Rhodesia for Copperbelt workings, and, as the Salisbury branches were dieselized, they appeared increasingly on the Bulawayo shunts, and the Redcliffe steelworks branch.

The final new Garratt class in Rhodesia was the 20th/20A class, different only in the diameter of the inner pony truck wheels – hardly sufficient to warrant reclassification. Sixty-one of these magnificent monsters were built in 1954, and were used mainly in Northern Rhodesia, where nearly three-quarters of the class was used for heavy coal and copper traffic. The general appearance, with sloping-sided front tanks, was in Beyer Peacock's final and finest tradition, and they rank with the classic designs used in East Africa (59 class), and New South Wales (AD 60 class). At first several weaknesses were found in the engine's boilers, but with these sorted out, the 20s settled down to do consistently good work with an impressively quiet competence. Dimensionally they are closely similar to SAR's GMAM class, also a 4-8-2+2-8-4. As late as 1979 the two classes on adjacent railways could be compared side by side on the same job between Wankie and Bulawayo, where the RR design was found to be more economical – with very similar tractive effort, boiler design, and identical bunker capacities, the GMAMs ran out of coal where the 20s got through.

268

267. **20A Class No. 722 roars out of Thomson Junction with an afternoon southbound coal train in July 1975. Entering the horseshoe at 'Christine's Curve', the Garratt is working hard on the climb from Tajintunda siding to the tunnel near the Baobab Hotel.**

268. **Early morning at Wankie Colliery, and coal is heaved upgrade to the washing and coking plants by a former SAR 16DA 4-6-2 and a 19D 4-8-2, banked by another 19D, in a most spectacular colliery operation.**

ZAMBIAN RAILWAYS

After the break-up of the Central African Federation, which included Nyasaland (now Malawi), the railways of both Zimbabwe (formerly Southern Rhodesia), and Zambia (Northern Rhodesia) were still physically united under what was termed the Unitary System. In spite of the ideological difference which steadily widened, Zambian Railways were entirely dependent upon Bulawayo works for repairs to its steam fleet, as no major workshops had been built in Northern Rhodesia since they were not necessary under the single railway system. The Zambian Railways ended up with 91 steam locomotives, divided as follows: 9A, 1; 9B, 3; 12th, 24; 15th, 5; 15A, 4; 16A, 11; 20th, 15; and 20A, 28.

As friction grew between Zambia and Rhodesia, following UDI, it was eventually decided to commission a new workshop at Kabwe (formerly Broken Hill) which was opened in 1971. Kabwe was fully equipped for steam overhauls, but Zambia was persuaded to go diesel, a decision which now places an increasingly heavy burden on Zambian Railways as purchasers of both fuel and spare parts. At present at least one authority within the country sees the logic in using coal from the new Mamba mine as fuel for steam locomotives, currently disused, but which can be rebuilt for further service, and some 20th Class have been returned to service and 10, 12R Class hired from SAR.

Zambesi Sawmills Railway

The realities of this railway apparently exceeded the imagination, according to those who have visited the system. Stretching north-west from Livingstone, a name in itself sufficient to evoke thoughts of the 'darkest Africa' of only a century ago, the ZSR was in fact a delightful, ramshackle line powered by the most incredible collection of second-hand motive power gleaned from

269

269. Zambesi Sawmills operated as varied a group of second-hand engines as could be found anywhere in southern Africa. There were four ex-RR 10th Class 4-8-2s, the first Mountains in Rhodesia, and in 1971 No. 156 arrived at Mulobezi with the overnight mixed train from Livingstone.
270. Zambian Railways 12th Class 4-8-2 No. 204, on hire to ZSR, heads the 09h00 mixed from Mulobezi to Livingstone, across the Sinde River bridge in January 1977.
271. At least eight former SAR and seven former RR 7th Class 4-8-0s were acquired from the 1920s to the early 1970s. Last was 1021, an SAR 7A, recently purchased when photographed leaving Mulobezi in 1971 with a train of empty timber wagons, heading for Buwe Pool working area in the Fakuda forest.

270

272. The only SAR 8th Class acquired arrived in 1971 (previously five slide-valve RR 8th Class 4-8-0s were purchased). Two years later, with blue paint already faded, she shunts passenger stock at Mulobezi.
273. Portrait of 1126 by Peter Bagshawe. Like other ZSR locomotives she retained her original number. Two SAR 13th Class 4-8-0T+Ts and a Malawi Railways G Class 2-8-2, No. 57, graced the roster of more than 30 steamers used over a 50-year period. Reportedly, steam is still in use.
274. Main-line steam in Zambia ended in 1973 though shunting and hauler work continued. In 1970 9B No. 84, one of the relatively few steam locomotives to receive ZR initials, shunted the Broken Hill Propriety Mine at Kabwe.
275. Four industrial ore railways operated on the Zambian Copperbelt. Nchanga Consolidated Copper Mines at Chingola ran two 1924 Hunslet 2-6-2Ts, and one 1947 Hunslet 0-8-0T. This 2-6-2T, No. 1, was stored in 1972 and though well painted, she probably never ran again.

271

272

273

274

275

276. **Heading a surprisingly short freight, an RR 15th Class Garratt leaves Lobatsi, Botswana, on a southbound international train to Mafeking.**

available sources in southern Africa. No recent first hand reports are available, but we understand that the last few operational 16As in Zambia have been drafted to work this line, now incorporated into ZR.

BOTSWANA

Botswana, the territory formerly known as Bechuanaland, reflects the name of the Tswana people who inhabit the south-eastern part of the country and form the majority of its population. For most of its existence the Bechualand Railway, already mentioned, has been operated by Rhodesia Railways. However, there are now plans to run the country's railways independently, and though no firm plans or dates have yet been established, 'Botswana Railways' have already purchased 47 coal wagons for internal use. Steam operation of the main line ceased in mid-1973, although for several years after this, RR maintained a locomotive stationed at Mafeking to shunt the yard at Lobatse; an instance, probably unique, of one country providing a locomotive stationed in a second country to carry out duties in yet a third country.

The future for steam in Botswana is slightly hopeful. The Selebi Pikwe mines in northern Botswana have offered their diesels for sale, and are actively implementing steam traction as the obvious answer, in a coal producing country, to the world oil situation. As a fairly small operation, under private management dedicated to profitable working, this is an obvious solution. Commercial considerations clearly favour the use by the Botswana Railway Corporation of local coal supplies for steam locomotion as the most economic means of hauling BRC's traffic.

Zimbabwe, the final phase

By the early 1970s, Rhodesia Railways was programmed to eliminate steam by 1980. 'Elimination of Steam' had already assumed an almost mystical significance among most of the world's railway managements and one could almost imagine that any heretical railway manager who defied the decree would be burned at the stake in the town centre of la Grange, Illinois, USA. However, Rhodesia was by then involved in the international sanctions imposed as result of UDI, and this gave RR an insight — yet to be revealed to many of the world's railways — into the true cost of diesel traction. The first oil crisis of 1973 alerted RR management to the realities of the situation, and by the time the Shah of Iran was deposed, RR was the only railway in the world with a policy geared to the non-availability of oil supplies. In reversing the policy of 'steam elimination', 87 RR Garratts are being rebuilt with modern features by RESSC - a firm which has set up a locomotive department, under an ex-RR engineer, to cope with this development. The locomotives comprise classes 14A, 15th/15A, 16A, and 20th/20A, and barely had the programme been started when two 19th class engines were being overhauled for Botswana — a move which could herald the return of steam in an area threatened by the world's oil crisis, yet where coal is still plentiful.

The thinking behind this is not difficult to comprehend. Diesel traction is currently about three times as efficient as existing steam, and where energy costs are at all comparable, is more economical. In southern Africa, however, diesel fuel costs are some 30 times the cost of coal, so that a diesel locomotive, even if three times as efficient as current steam, costs *ten* times as much as steam to fuel. Under these circumstances, it is no wonder that some railway administrations are returning to steam, although there are several organizations, afflicted with bureaucratic inertia, who cannot or will not see the logic of steam renaissance.

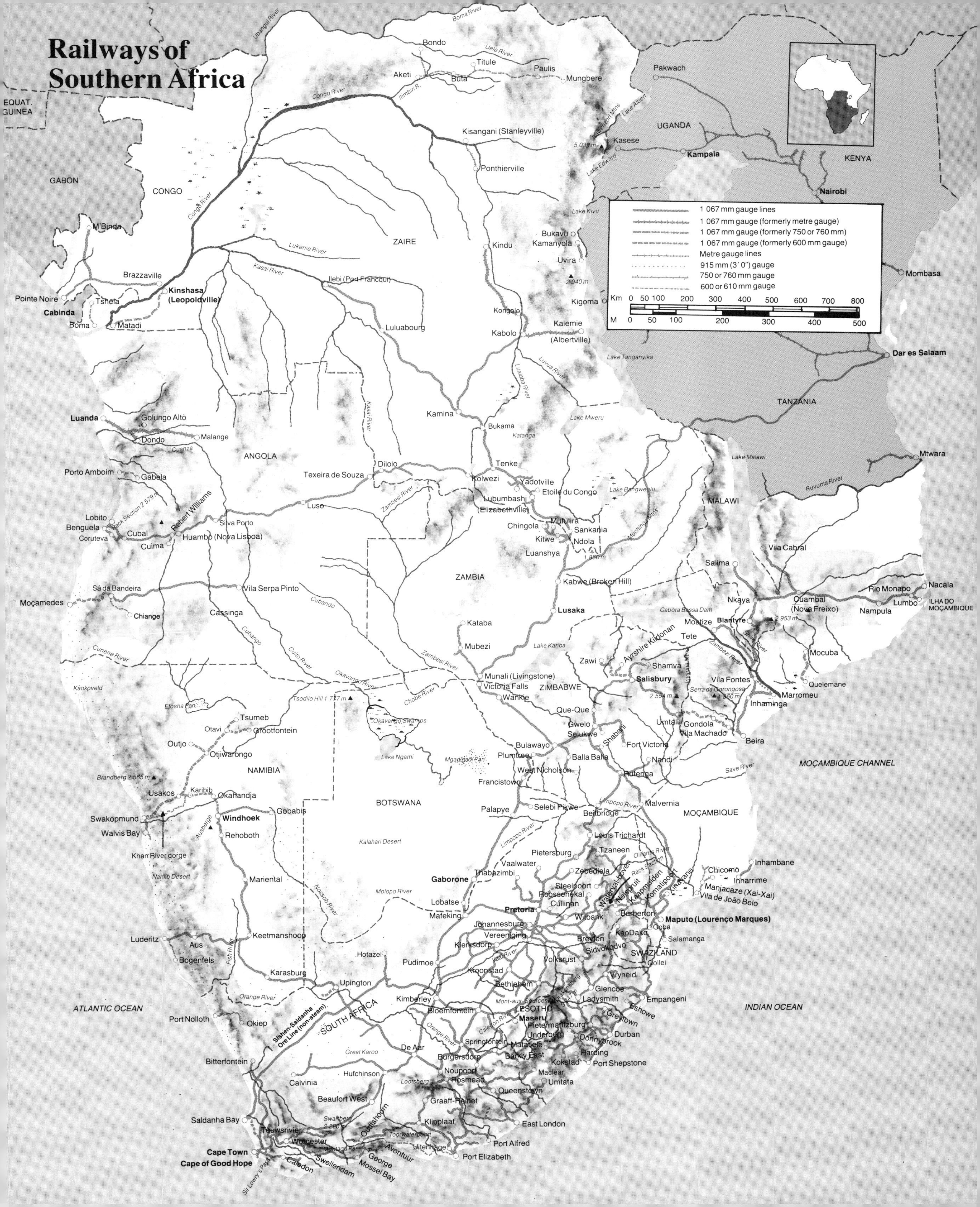

Railways of Southern Africa
1 067 mm gauge lines
1 067 mm gauge (formerly metre gauge)
1 067 mm gauge (formerly 750 or 760 mm)
1 067 mm gauge (formerly 600 mm gauge)
Metre gauge lines
915 mm (3′ 0″) gauge
750 or 760 mm gauge
600 or 610 mm gauge
Km 0 50 100 200 300 400 500 600 700 800
M 0 50 100 200 300 400 500
EQUAT. GUINEA
GABON
CONGO
ZAIRE
UGANDA
KENYA
TANZANIA
ANGOLA
ZAMBIA
MALAWI
ZIMBABWE
NAMIBIA
BOTSWANA
MOÇAMBIQUE
SWAZILAND
LESOTHO
SOUTH AFRICA
ATLANTIC OCEAN
INDIAN OCEAN
MOÇAMBIQUE CHANNEL
Ubangui River
Boma River
Bondo
Uele River
Titule
Paulis
Mungbere
Aketi
Buta
Itimbiri R.
Congo River
Pakwach
Lake Albert
Kisangani (Stanleyville)
Kasese
Kampala
Lake Edward
Ponthierville
Nairobi
M'Binda
Lake Kivu
Lukenie River
Bukavu
Kamanyola
Kindu
Uvira
Brazzaville
Kasai River
Ilebi (Port Francqui)
Mombasa
Pointe Noire
Tshela
Kinshasa (Leopoldville)
Cabinda
Kigoma
Boma
Matadi
Kongolo
Luluabourg
Kalemie (Albertville)
Kabolo
Dar es Salaam
Lake Tanganyika
Luvua River
Lualaba River
Kasai River
Kamina
Lake Mweru
Luanda
Golungo Alto
Bukama
Katanga
Dondo
Malange
Cuanza
Lake Malawi
Mtwara
Porto Amboim
Dilolo
Tenke
Gabela
Teixeira de Souza
Kolwezi
Yadotville
Ruvuma River
Etoile du Congo
Lake Bangweulu
Lubumbashi (Elizabethville)
Luso
Zambesi River
Robert Williams
Lobito
Benguela
Coruteva
Rack Section
Cubal
Silva Porto
Mufulira
Chingola
Sankania
Cuima
Huambo (Nova Lisboa)
Kitwe
Ndola
Muchinga Mtns
Luanshya
Vila Cabral
Salima
Sá da Bandeira
Vila Serpa Pinto
Kabwe (Broken Hill)
Nacala
Rio Monapo
Moçamedes
Cubando
Nkaya
Quambal (Nova Freixo)
Lumbo
ILHA DO MOÇAMBIQUE
Cassinga
Lusaka
Cabora Bassa Dam
Nampula
Chiange
Kataba
Moatize
Blantyre
Cubango
Tete
Mubezi
Lake Kariba
Ayrshire Kildonan
Shire River
Cunene River
Culto River
Zambesi River
Mocuba
Zawi
Shamva
Zambezi River
Kaokoveld
Okavango River
Munali (Livingstone)
Salisbury
Vila Fontes
Victoria Falls
Quelemane
Chobe River
Serra da Gorongosa
Marromeu
Tsodilo Hill
Wankie
Etosha Pan
Inhaminga
Okavango Swamps
Que-Que
Tsumeb
Gwelo
Umtali
Gondola
Otavi
Grootfontein
Selukwe
Shabani
Vila Machado
Fort Victoria
Beira
Outjo
Bulawayo
Lake Ngami
Mgadikgadi Pan
Plumtree
Balla Balla
Otjiwarongo
Nandi
West Nicholson
Save River
NAMIBIA
Brandberg
Francistown
Rutenga
Usakos
Karibib
Okahandja
Swakopmund
Windhoek
Gobabis
Palapye
Selebi Pikwe
Beitbridge
Limpopo River
Malvernia
Walvis Bay
Ausberge
Rehoboth
Limpopo River
Kalahari Desert
Louis Trichardt
Khan River gorge
Pietersburg
Tzaneen
Olifants River
Vaalwater
Namib Desert
Zebediela
Rack Section
Inhambane
Chicomo
Inharrime
Mariental
Gaborone
Thabazimbi
Steelpoort
Roossenekal
Waterval-boven
Nelspruit
Kaapmuiden
Komatipoort
Xinavane
Manjacaze (Xai-Xai)
Vila de João Belo
Molopo River
Nossob River
Lobatse
Pretoria
Cullinan
Mafeking
Witbank
Barberton
Maputo (Lourenço Marques)
Johannesburg
Goba
Luderitz
Keetmanshoop
Vereeniging
Breyten
KaoDake
Salamanga
Aus
Klerksdorp
Sidvokodvo
Fish River
Bogenfels
Hotazel
Pudimoe
Vaal River
Volksrust
Gollel
Karasburg
Kroonstad
Vryheid
Upington
Bethlehem
Glencoe
Orange River
Kimberley
Ladysmith
Empangeni
Bloemfontein
Mont-aux-Sources
Eshowe
Port Nolloth
Okiep
Sishen-Saldanha Ore Line (non-steam)
Maseru
Greytown
Pietermaritzburg
Orange River
Caledon River
Underberg
Durban
Springfontein
Donnybrook
Great Karoo
De Aar
Matatiele
Harding
Burgersdorp
Barkly East
Bitterfontein
Kokstad
Port Shepstone
Noupoort
Maclear
Calvinia
Hufchinson
Lootsberg
Rosmead
Umtata
Queenstown
Beaufort West
Oudtshoorn
Graaff-Reinet
Saldanha Bay
Swartberg
Klipplaat
East London
Touwsrivier
Toorwaterpoort
Worcester
Montagu Pass
Avontuur
Port Alfred
Cape Town
Uitenhage
Cape of Good Hope
Caledon
Swellendam
George
Mossel Bay
Port Elizabeth
Sir Lowry's Pass

11 NAMIBIA and THE REPUBLIC OF SOUTH AFRICA

NAMIB STEAM

The problems in operating a steam railway through the desert are many and varied. The Germans who built the Otavi railway at the turn of the century – 700 km of 600 mm-gauge metals to serve the copper mines at Tsumeb – could vouch for this. Sand drifting across the track, water scarcity, rails (and tempers) buckling in the heat – they had expected these hazards.

But the most improbable and bizarre manifested itself on a scorching day in the 1950s as a little train trekked along the edge of the Namib desert. Ahead of the caboose were five combination day-and-sleeping coaches; seven wagons loaded with brightly-painted trailers, cages and other paraphernalia; and four covered vans housing performing animals. Immediately behind the engine a high-sided open gondola accommodated two elephants. The circus was on its way to Tsumeb.

The crew of the tiny Kalahari Mikado were battling in the midday heat to coax her up the grade from Otjikango to Platveld when, about two kilometres from the summit, the left-hand injector knocked off without warning; nor would the right-hand one work. The fireman scrambled up the coal pile to check the tank, only to see the last of his precious water siphoned into an elephant's trunk and ejected in a cooling spray over the owner of the trunk and his mate.

The crew could do nothing but cut off the load and run for water. The circus opened late in Tsumeb that night.

Narrow-gauge steam no longer operates in Namibia, the section from Walvis Bay to Usakos having been widened during World War I, and that from Usakos to Tsumeb having been widened and dieselized along with the branches between 1958-61.

In 1884 the Germans established a protectorate between the Cunene and Orange rivers and German South West Africa came into being. The following year a private company, the Deutsche Koloniale Gesellschaft, was founded to exploit the minerals known to exist in the territory.

Although numerous surveys were made and two short lines built – at Cape Cross to harvest guano deposits there, and at the British enclave of Walvis Bay – no railway of any significance was built until 1897, five years after the protectorate had become a colony. It was only then that the German government began a 600 mm-gauge military railway eastwards from Swakopmund for their campaign against the Herero.

Lightweight Feldbahn equipment was used with the track prefabricated from 9-kg rails and metal sleepers laid down on rudimentary formation. Most of the route followed the Swakopmund-Windhoek wagon road. It reached Windhoek in 1902, and soon supported a creditable two-day passenger service over the 382 km – with an overnight layover at Karibib, location of the main workshops.

For its first few years the railway was primarily a supply line for the German forces fully occupied fighting the Herero. However, by 1907 conditions were sufficiently stable for it to be handed to civilian control, and thereafter the 'State Northern Railway' operated commercially.

The line's profile was rugged, climbing from sea level to 1 654 m at Windhoek, with several minor summits on the way. The major obstacle was the Khan River gorge, negotiated by 1-in-19 grades on the eastern bank.

Initial main-line power consisted of double 0-6-0Ts coupled back to back and operated by one crew. These 'Zwilling' (twin) locomotives could be operated singly, though in this form they were not very effective – particularly out of the Khan gorge where even twins could manage only a 20 gross tonne trailing load. Operations up the sustained 1-in-19 grade eastbound required two and sometimes three 'Zwillinge' and the authorities decided that eight-coupled power was better suited to the terrain. The first of the classic Feldbahn 0-8-0Ts were delivered by Krauss in 1901 and this type – later produced by the thousand for Germany's World War I campaigns – also became popular on the State Northern, where 20 were in service by 1905. This was the peak year for motive power, with no less than 73 twin units also on the roster.

The presence of extensive copper deposits around Tsumeb had been

277

278

279

known for more than 60 years when, in 1900, the Otavi Minen und Eisenbahn Gesellschaft (OMEG) was formed to exploit them and export the ore via a new 600-mm railway. This was to be rather more substantially built than the State Northern (SN).

The contract was awarded to Arthur Koppel & Co., part of the famous locomotive-building firm. Begun at Swakopmund in 1903, the first 178 km roughly paralleled the SN, though on a more satisfactory ruling grade of 1-in-66. At Km 178 only 13 km separated Onquati, on the OMEG, from Karibib on the SN. By 1906 the OMEG had reached Tsumeb and the first trainloads of ore headed for the coast.

The first motive power on the Otavi Railway was not provided by Jung and by Henschel – presumably because Orenstein & Koppel had their order books full in those prosperous times. Of 25 0-6-2Ts built only 23 were delivered, two being lost at sea. Some of these successful engines survived until the end of South West African narrow gauge, and at least one, complete with shellhole in the cab side-sheet, gave more than 40 years' service on the Mount Edgecombe sugar estate in Natal, where it was sent after World War I.

Like the SN, the OMEG sited its main workshops inland, but at Usakos, 150 km away from the corrosive atmosphere at the coast. As at Karibib, extensive repairs, such as replacement of fireboxes – a common operation where only hard water was available – and even construction of new rolling stock was carried out.

The last skirmishes with the Herero, during 1906, taxed the SN fully. At least one OMEG Jung 0-6-2T was commandeered by the military and pressed into helper service out of the Khan gorge.

Moir and Crittenden, in *Namib Narrow Gauge* describe train workings from Khan: An eleven-car train would arrive at Khan hauled by a Zwilling double 0-6-0T. Here the train would split up, the first two wagons being worked on up the 1-in-19 to Welwitsch by the Zwilling. Doubleheaded Feldbahn 0-8-0Ts would then take out the next five wagons, followed by an OMEG Jung with the remaining four. At Welwitsch the whole caravan would be re-assembled and worked onwards by the Zwilling.

By 1907 the OMEG had acquired 41 engines, all from Jung and Henschel, of which 38 were 0-6-2Ts and three were ultralight 0-6-0Ts for shunting on the pier at Swakopmund and on the mines.

For five years South West Africa enjoyed virtually duplicated rail facilities for the first 180 km inland from Swakopmund. At last, in 1910, agreement was reached between OMEG and the SN that traffic up from the coast would be handled by the Otavi Railway, utilizing a link between the two lines from Karibib to Onquati. The old line through the Khan gorge was reduced to a skeleton service of a fortnightly train serving the small communities en route.

Whereas the SN had lost all its traffic west of Karibib, the OMEG saw unprecedented increases, not only in general merchandise and passengers, but also in ore traffic. The little 0-6-2Ts began to lose their grip, and orders were placed with Henschel for three engines of greatly increased power. The result was OMEG class HD, one of the classic designs for 600 mm gauge, which would be re-ordered many times, with only detail modifications, for nearly 50 years. The provision on such a narrow gauge of a 60 tonne superheated 2-8-2 with 8 500 kg (18 800 lbs) of tractive effort was no mean feat in 1911.

Many surplus SN locomotives were sold to the flourishing network of 600 mm lines, built since 1908 to serve the newly-discovered diamond fields south of Luderitz. At their peak these lines totalled more than 200 km, but were soon to dispense with the services of their steam locomotives, as water for them was almost unobtainable.

The first Cape-gauge line in the territory was started in 1905 from Luderitz, reaching Keetmanshoop on the far side of the Aus mountains in 1908, the year in which two branches were started southwards off this line. One, the 600-mm feeder to the diamond fields, began at Kolmanskop, heading for Elizabeth Bay and Bogenfels, 120 km away. A maze of sub-branches of this

277. **Forerunner of a class produced with detail modifications for nearly 50 years, the OMEG Class FD 2-8-2 of 1911.**
278. **An impressive line up at Usakos shed in 1955 showing from the left, Classes NG5, 10, 15, 5 and 10.**
279. **Zwilling 167A and B of the State Northern Railway on shed. Note the auxiliary water tender.**
280. **An ex-OMEG Jung 0-6-2T with auxiliary tender heads a typical rake of Otavi stock, including a passenger brake from the Avontuur branch.**

280

line served the actual diggings. The other branch stretched 160 km from Seeheim to Karasburg and eventually became part of the main line to De Aar.

After reaching Keetmanshoop there was a lull of some two years before railway building recommenced towards Windhoek, which by 1912 had its 'broad' gauge connection with the sea. Construction gangs then proceeded northwards to widen the 600-mm to Karibib, completing the job in 1914.

Five classes were used on the 1 067 mm gauge up to this time. To work the steep climb inland over the Ausberg from Luderitz, six 0-10-0s were supplied by Henschel in 1910. For the more 'level' territory inland, Orenstein & Koppel supplied eight Von Borries two-cylinder compound 2-8-0Ts and eight slightly larger two-cylinder simple 2-8-0s. In 1911 O & K supplied a further 15 of a simple version of the 2-8-0T and auxiliary tenders for the long, waterless sections, and two small 0-6-0Ts for shunting.

German railway achievement in South West Africa reached a climax in 1914. About 1 250 km of 1 067 mm-gauge railway were already in operation and it had been decided to widen the 600 mm-gauge line from Swakopmund to Omaruru and strengthen the section from there to Tsumeb, and some interesting 2-10-2s were in the pipeline. The Otavi Railway was flourishing, with as many as 47 trains a week, and planned improvements to the Tsumeb passenger services were so advanced that two chic little Pacifics, utilizing the 2-8-2 boiler, had been built by Henschel.

Then came World War I. The Pacifics were never delivered; the 2-10-2s were never built; and the Germans of South West Africa prepared to defend their adopted land without much help from their fatherland.

When the South African forces landed at Walvis Bay their first objective was Swakopmund, and to cross the dune country a railway was essential.

They chose 1 067 mm gauge, for the State Northern had already been widened to Karibib from Windhoek, and this meant that through iron on the Cape gauge existed all the way from Karasburg. It only remained for engineers to finish their extension of the De Aar-Upington line to Karasburg, and the Windhoek-Karibib line to Swakopmund to create through-rail communication on one gauge between Walvis Bay and South Africa, and German South West Africa could be conquered.

To avoid the Namib's giant shifting dunes, the sappers laid the line from Walvis Bay along the beach, just above the high water mark, but drifting sand was a problem and eventually they covered trouble spots with tarpaulins. When they reached Swakopmund the town was empty; all vital installations, including the water-softening plant, had been destroyed; and the rails of the Otavi line had been mined at irregular intervals. Surprisingly, the old SN line through the Khan gorge had been left mostly intact and was soon humming with five small engines, hurriedly shipped from the Union by the South African army. Four came from the Reef mines and one from the Esperanza Sugar Estate in Natal. These five and two old SN engines which had been resuscitated were all that could be used on the light track to support the march on Karibib.

Meanwhile, work on converting the Otavi Railway to 1 067 mm gauge went ahead quickly. All facilities had been destroyed by the retreating Germans and numerous mines laid under rail joints slowed the work, but the link-up was eventually made at Karibib in August 1915. By this time the eastern engineers had thrust through from Karasburg and, for the first time in Africa, a transcontinental link had been created – from Walvis Bay to Durban, a distance of more than 3 000 km.

When the last German resistance was quashed at Otavi, the major task of rebuilding the Otavi Railway began. All bridges had been destroyed; all locomotives immobilized; the workshops at Karibib and Usakos sabotaged

283

beyond repair; and none of the vital water-softening plants remained intact. Though, for a while, the extensive SN shops at Karibib were reinstated, the original SN Swakopmund line was abandoned in 1916, and all narrow- and Cape-gauge repairs were concentrated in the OMEG shops at Usakos. The Karibib-Onquati line was uplifted, making Usakos the new southern terminus of the Otavi Railway.

With almost all Otavi locomotives out of action, the army commandeered 14 SAR narrow-gauge locomotives of different types. The 13 repairable OMEG engines were put back into traffic as quickly as possible by the workshops, which also had hundreds of items of rolling stock to repair – and even built a 600 mm-gauge dining-car, which became highly popular on the Tsumeb run in later years.

For the 1 067 mm gauge the army brought 45 SAR engines, of these 28 were class 7 – a type which was to make South West Africa its home, handling nearly all road operations until the arrival 35 years later of the class 24 2-8-4s. For work close to the front, five class 05 engines were armour-plated and given the names 'Trafalgar', 'Scot', 'Erin', 'Karoo' and 'Skrikmaker'.

Things soon settled down after the South West African campaign. OMEG continued to operate the mines, though they had lost control of their railway. Coal was no longer imported from Germany, instead it came by rail from Witbank – a haul of 3 000 km to the locomotive depot at Tsumeb.

Though some good engines had been introduced on the Otavi line by the army, none could compare with the three Henschel HD class 2-8-2s of 1911. Six of the same type were ordered from the same firm and were delivered in 1922. When the SAR formally took over South West Africa's railways in 1923 these engines became class NG5.

In 1927 a SAR class NGG12 Garratt was transferred to the Otavi line; though slow, it was otherwise successful and two more Garratts were brought in, another NGG12 and the solitary NGG14. These were the only three articulateds to run in South West Africa where they continued until the arrival of the class NG15s from Henschel in 1931, when they were returned to South Africa.

The NG15s are direct descendants of the HDs of 1911, the only obvious difference being the moving of the pony truck to its customary position in front of the cylinders. So successful was this type that repeat orders were placed with various builders at intervals until 1958, 47 years after the original HDs went into service.

When the Otavi Railway was broadened all 21 of the NG15s were transferred to the Avontuur line where they were still in service in 1980. On a gauge of only 600 mm, these engines could run more than 5 000 km a month, so that those in the 1931 batch are now approaching 3 000 000 km – a record unlikely to be equalled by their new diesel-electric rivals.

281. **Bags of guano for the farmers of the Highveld have been off-loaded from the sail-barges at Luderitz pier. Soon the old 7th Class will shuffle these wagons into a train and battle up to the 1 500 m summit at Ausnek.**

282. **A full rake of main-line stock, complete with twin dining car, forms the Walvis Bay to Windhoek mail, here drawn by a 7th Class locomotive, south of Okahandja. George Kempis, who took this picture as late as 1955, records that one week later the train went over to 24 Class haulage. This was most certainly the very last main-line train regularly hauled by southern Africa's 'maid-of-all-work' 19th century locomotive, the famous 7th Class.**

283. **An O&K two-cylinder compound 2-8-0T of the old State Southern Railway, thought to be at Luderitz about 1911.**

After World War II, the OMEG was sold, lock, stock and barrel, to the Tsumeb Corporation. The mines, closed since 1940, were re-opened in 1947 and operations on the company's private tracks around Tsumeb were handled by four ex-OMEG machines – three Henschels and one Jung 0-6-2T – which ran until 1958. They were the last of the original OMEG locomotives in service.

Traffic on the Otavi line increased greatly during the 1950s and, in spite of support from the ten NG15s from Franco-Belge which arrived in 1951-52, it became apparent that before long capacity would be limited by the gauge. The Tsumeb Corporation could not pin down the SAR to an actual date for gauge conversion, and the traffic position was becoming so alarming that, as an interim measure, the TC ordered five more NG15s from Henschel and seven NGG16s from Beyer Peacock, on condition that the SAR agreed to buy them when they widened the gauge.

No sooner were these engines ordered than the SAR decided that the gauge should be widened as soon as possible. In July 1960 work was completed to Tsumeb; cleaning up and track-lifting took another year, and in September 1961 NG15 No. 40 was steamed for the last time at Usakos, bringing down the curtain on 65 years of narrow-gauge operations in South West Africa – and without giving the newly-delivered Garratts a chance to play their part.

Over the years more and more class 7 locomotives had been drafted in to the desert operation. They had been improved by superheating but were approaching 60 years of age – and it showed. The class 24 had been introduced to South West Africa from 1951 expressly to release the aging 7s, but the latter were never entirely dispensable.

By the mid-1950s increasing traffic and diminishing water supplies presented problems which clearly would take a lot of solving with steam traction.

284. **The mail train to South Africa required at least two, and sometimes three, engines to surmount the Kruin Bank, southbound from Windhoek. Here a pair of 24s tackle the climb in 1955.**

285. **Seen by many and photographed by few. Primitive dirt roads restricted road transport in South West Africa until the 1960s, but in 1957 this 19D with the southbound mail, was encountered crossing a stark desert landscape near Karasburg.**

Condensing locomotives were considered, but at that time problems were being experienced with the 25s; there was the cost of railing coal from 3 000 km away; and, all things considered, the decision to dieselize was inevitable.

By 1962 all steam operation west of De Aar had ceased, except for a brief revival as far as Prieska during a diesel shortage in the early seventies. The released class 24s were all dispersed elsewhere on the SAR, but most of the class 7s went to scrap though some survived on various branch lines for another decade.

CAPE-GAUGE STEAM

'Once we worked to the nearest millimetre; now it's one tenth of a millimetre.' The locomotive foreman at Beaconsfield referred with mixed pride and irritation to the exhaust nozzle settings on locomotive No. 2644, once an orthodox Stephensonian 19D, being transformed under his supervision to a sparkling performer equipped with Lempor exhaust and gas-producer firebox. This low-cost conversion led to the large increase in efficiency forecast by the designer, David Wardale, and his mentor in Patagonia, L.D. Porta.

Just when it seemed that the steam locomotive would soon become extinct, Porta devised a mutation which should allow it to survive well into the next century. His gas-producer combustion system and other major improvements have removed the two points which were (as Baron Vuillet put it) steam's Achilles' heel: high maintenance costs, and decreased boiler efficiency at high combustion rates.

South Africa, blessed with huge coal reserves, is an ideal breeding ground for the new strain, for gas-producer grates thrive on low quality coal which is not in great demand elsewhere. Already conventional steam is more economical than diesel; and it is anticipated that gas-producers will eventually prove more economical than even electric traction.

The story of steam in South Africa goes back more than 120 years, to a September day in 1859 when a construction locomotive for the contractors to the Cape Town Railway and Dock Company was offloaded in pieces from the brig *Charles,* assembled on wooden rails on the dock and trundled manually up to the square known as the Grand Parade. The contractors were E. and J. Pickering, of North African fame, and their tiny standard-gauge well tank was the forerunner of the largest locomotive fleet in Africa.

The chief engineer, W.G. Brounger's planning of the Cape Town-Wellington railway, which ran via Stellenbosch, set standards of civil engineering unequalled elsewhere on the railways of South Africa for many years.

To work the Wellington service, eight 0-4-2s were supplied by R. & W. Hawthorn and Co. in 1860, but owing to tardy tracklaying by Pickerings, commercial services began over three kilometres of line only in 1861, by which time the laurels for opening the first public railway in South Africa had been snatched by Natal.

In June 1860 the first public train had set out on its 3 km journey from Durban to the Point. For several years the only mechanical motive power on the line was 'Natal' a minute standard-gauge well tank by Robert Legg. Campbell, in *The Birth and Development of the Natal Railways,* tells how when she was unable to work 'she had perforce to suffer the indignity of seeing her precious train being hauled along the rails by animal power'. In 1867 the Point railway was extended 5 km from Durban to Umgeni where stone for the harbour works was quarried.

It was an altogether less substantial affair than Brounger's railway to Wellington, as evidenced by the following quotation of Campbell's from the *Natal Mercury* of 1875: 'Owing to the heavy rains, the trains are discontinued

pro tem on both Point and Umgeni lines. Before resuming traffic the steam whistle will be sounded for five consecutive minutes.'

The 1860s saw completion of the 96-km Cape Town-Wellington railway and a 10-km branch through the southern suburbs of Cape Town to Wynberg, but the interior of South Africa remained unattractive to the rail barons; the railways remained stubbornly at their standard-gauge terminals of Wellington, Wynberg and Umgeni.

However, with the upsurge of the great diamond rush in 1869, all this changed. The diggings centred on the rich pipes at Kimberley, and transport from the railhead 1 000 km away was at a premium, in spite of the gruelling 16-day stage-coach journey through the Karoo. Yet politics and the gauge controversy delayed the start of railway construction beyond Wellington.

As the enormous amount of capital needed could not be raised by private enterprise, the Cape Government decided to undertake the extension and operation of railways. On January 1, 1873, it took over the Cape Town Railway and Dock Company, retaining the able Mr Brounger as chief engineer.

It was now that the gauge – eventually to be adopted from the Cape to the Congo and which was to become known as 'Cape gauge' – was decided on. The struggling colonial government wanted the extension to be built as cheaply as possible. In 1869 R. Thomas Hall had started to build a 2′ 6″ (760 mm) gauge line for the Cape Copper Company from O'Kiep to Port Nolloth. In such dry, riverless country it was possible to build a very cheap railway, without bridges, earthworks or ballast. Its low cost convinced many politicians that narrow-gauge railways were much cheaper to build. Hall's favoured 2′ 6″-gauge was rejected by only a small majority. The persuasive Mr Hall had stated with Victorian assurance: 'Nothing can be done on a 4′ 8″ or 3′ 6″ gauge which cannot be done on a 2′ 6″ gauge.' A compromise gauge of 3 ′6″ (1 067 mm) was chosen and Natal wisely followed suit, deciding that all new construction should be on Cape gauge; after construction of the line to Pietermaritzburg was started, standard gauge was eliminated at Durban.

The Cape Government Railway (CGR) began their extension beyond Wellington in 1874. To avoid trans-shipping, a third rail was laid inside the standard gauge back to Cape Town. By 1881 the outside rail had been removed, ending the standard-gauge era in South Africa. At its peak there had not been much more than 100 km of track and D.F. Holland lists only 17 standard-gauge locomotives as ever having existed in South Africa. The first, Pickering's construction engine, is now on show in Cape Town's station.

Only in 1885 were the Kimberley diggers able to celebrate the arrival of the first train from Cape Town; but this was not the only line to be built in the preceding decade. The discovery of diamonds so increased the prosperity of the colony that the government could justify and finance a series of lines inland from the coast. By 1884 the Western and Midland systems had met at Brounger junction (now De Aar), thus providing through-rail communication between Cape Town and Port Elizabeth. Also by this time the main line from East London had surmounted the Stormberg at an altitude of 1 720 m.

In 1872 the Natal Railway Company declared its first dividend, and colonists everywhere were clamouring for lines to their areas. At the insistence of the British government, in 1877 the existing line and responsibility for new construction was taken over by the Natal government. The Natal Government Railways (NGR) started simultaneously to build lines from Durban to Pietermaritzburg, and along the north and south coasts. On ruling grades of 1-in-30, with uncompensated for curves of 100 m radius, the capital was reached in 1879, and a daily service of one train in each direction was introduced. Its 6 hour 14 min schedule was 54 minutes slower than the best times offered by the rival stage-coach company.

The most important objective for the NGR was the large coal measures of Newcastle and Dundee. Good steam coal had been known to exist in Natal since 1838, and the railways and an increasing number of steamships at Durban needed it. Construction beyond Pietermaritzburg proceeded via Ladysmith until Newcastle was reached in 1890, enabling NGR, in 1891, to be the first to reach the Transvaal border. In 1892 the NGR completed the first section of their main line from Ladysmith to the Orange Free State, crossing the Drakensberg via Van Reenen's Pass to Harrismith 40 km away.

In the Transvaal, both President Kruger and his predecessor were aware of the benefits an outlet to the sea, independent of the British colonies, would bring.

As early as 1874, R. Thomas Hall was commissioned to survey a route to the Portuguese port of Lourenço Marques on Delagoa Bay. Money was raised in Europe and a railway tax was imposed in 1876, but sufficient funds could not be raised and the scheme lay dormant until the discovery of gold on the Witwatersrand in 1886, and the opening of the first coal mine in the Transvaal at Boksburg in 1887, gave Kruger his three prerequisites for a railway – gold to finance it and water and coal for its locomotives.

In 1887 the Netherlands South African Railway Company (NZASM) was granted the concession to build and operate railways in the South African Republic and agreement was reached that the Portuguese government's concessionaire would build a line from Delagoa Bay to the Transvaal border while NZASM would construct the section to Pretoria and Johannesburg.

Nevertheless Lourenço Marques was not the first port to be linked with the Transvaal goldfields. Kruger had at first refused the British colonial railways permission to cross his territory until the Delagoa Bay line was completed. Then NZASM ran into financial difficulties and Rhodes's Cape government lent them £1 million on condition that the CGR was allowed to cross the Vaal. It did so in 1892, connecting with the NZASM line to the Witwatersrand. Earlier in 1892 the main line from East London was connected with the Orange Free State main line at Springfontein, giving the Rand access to three Cape ports.

By this time hundreds of people had already died building the Delagoa Bay line across the tsetse fly and malaria ridden lowveld. George Pauling, the active partner of the British firm of James Butler and Co., was entrusted with one of the most difficult stretches, in Crocodile Poort, and the line eventually reached the fever-free highveld by means of a rack section up the escarpment at the head of the Elands River valley.

While Pauling and his men were sweating it out in Crocodile Poort other employees quietly completed the first public railway in the Transvaal. Called the 'Rand Tram' to mollify anti-railway voters, it was a 'tram' in name only for, though passengers were carried, its chief business was carting coal from Boksburg mines to the Reef. In 1890, 25 km were opened and it was extended to Krugersdorp in 1891.

Johannesburg and Pretoria were linked with Lourenço Marques in 1894 and a year later the NZASM joined up with the NGR at Charlestown to complete South Africa's main-line network. From 1895 until 1970 only secondary trunk routes and branch lines were added – though what would become a trunk route to Rhodesia was started by the Pretoria-Pietersburg Railway Company, reaching Pietersburg in 1899.

As the century ended the South African Republic had 1 420 km of railway; Natal, 800 km; the Orange Free State, 710 km (taken over from the CGR in 1897); and the Cape Colony, 4 000 km.

286. **The doyen of South Africa's locomotive enthusiasts, D.F. Holland, poses next to the country's first locomotive, the Pickering contractors' engine, built by Hawthorn & Co in 1859, at Salt River works in 1929.**

287 288

287. **An old CGR 4th Class 4-6-0 halted on the Hex River Pass about 1896.**
288. **The engines which had a most profound effect on locomotive development in South Africa were the six Atlantics supplied by Baldwin in 1897.**
289. **A Reid 4-10-2T heads for the Highveld with the Durban-Johannesburg corridor express about 1904.**
290. **One of the 8th Class ordered by the Imperial Military Railways in 1902 (and named 'Joseph Chamberlain' after the Colonial Secretary during the war) poses alongside a ZASM 14-tonner.**

Inevitably, the character of these lines influenced the evolution of their rolling stock. Trained British engineers and draughtsmen developed indigenous designs for locomotives, goods wagons and passenger coaches suited to African conditions. Derivations spread through the Cape-gauge systems and much of their inspiration rubbed off on the metre-gauge East African Railways as well.

Topography was an important influence on locomotive design. Ten or even 12 coupled wheels would have been ideal as all the main lines from the coast use long, fierce gradients to reach the 1 800-2 000 m elevation of the highveld; but the cheaply-laid track, with minimum earthworks and maximum curvature, restricted the locomotive engineers, who rarely ventured beyond eight coupled wheels. Instead, when more adhesion was needed they resorted to articulation. The severe curvature of the lines led to the early adoption of a four-wheeled leading bogie, while the generally high ash content of local coal meant big grates over a trailing two-or four-wheeled truck.

The CGR appointed Michael Stephens as their first locomotive superintendent. He kept abreast of locomotive developments and incorporated the best practice of other railways in his own designs.

A succession of small factory-designed engines preceded the introduction of Stephens's 4th class 4-6-0T + T, the first locomotives designed in detail at the Cape. From 1880 onwards 92 were built by Stephenson and by Neilson. Though more powerful than their predecessors, from which they had obviously evolved, these engines still incorporated some of the unsuitable features of the previous 'off the shelf' classes.

Tests with Syphergat coal on the Eastern main line led to the introduction of four experimental 4th class in 1884. They had practically double the grate area – 1,75 m^2 (18,25 ft^2) – and 40% more firebox heating-surface. These tests were so successful that all future CGR engines were designed with large fireboxes, grates and smokeboxes.

The 4-6-0 arrangement was already establishing itself internationally by the time Stephens's Cape 5th class arrived in 1890. All 50, built by Dübs, were highly successful, most giving over 50 years' service. This class spawned the succession of fine designs which the CGR contributed to the SAR at Union.

Reports of the success of the eight-coupled Dübs A tanks on the NGR led Stephens to Natal in 1890. Returning, he instructed his superintendent at Salt River Works, H.M. Beatty, to prepare plans and specifications for an eight-coupled engine for the CGR – and one of Africa's legendary locomotives, the Cape 7th class, was born. In various guises these 4-8-0s were to be found throughout Africa south of the Sahara. They even crossed the seas to Spain and Australia. Many gave more than 70 years' service and two still work on the colliery railways at Witbank.

Dübs delivered the first six in 1892 and from then on 72 more by Dübs, Sharp Stewart and Neilson Reid were placed in service. Each subsequent batch incorporated minor improvements, such as slightly increased heating-surface, eight-wheeled tenders and four-window cabs.

Another famous Beatty design, the CGR 6th class 4-6-0, was introduced in 1893. Where the 7th class were primarily goods engines, the 6th class were passenger and mixed-traffic locomotives. Though less widespread, they saw service throughout southern Africa, as well as in the Sudan. Altogether 229 of the 4-6-0 6th class and two 2-6-2s and eight 2-6-4s with the same classification were placed on the CGR until 1904. The first 180 had only minor variations, later models had important changes such as bar frames and wide fireboxes, the latter necessitating the altered wheel-arrangements.

When Stephens retired in 1895 he was succeeded by Beatty who was faced immediately with a locomotive shortage for the newly-opened Kimberley-Mafeking section. He was thwarted by the same 1897-98 British strike which forced other railways to place orders with American builders.

Fortunately, Baldwin was able to supply Beatty with six ready-built engines in 1897. These first and only 4-4-2s to run in South Africa were to a design which Baldwin had prepared for Japan and featured bar frames, with the classic arrangement for an Atlantic of wide firebox over trailing bissel. The British strike thus had a profound effect on the evolution of the locomotive in South Africa. Beatty was so impressed by these free-steaming Atlantics with grates larger than on any other existing South African locomotive, and rugged bar frames which proved ideal for colonial track, that he adopted both and they became standard on the CGR and, later, the SAR.

On the NGR the service between Durban and Pietermaritzburg commenced with various Beyer Peacock 2-6-0Ts, but the chief locomotive superintendent, W. Milne, soon realized that a four-wheeled leading bogie was needed on a line with 100-m-radius reverse curves. Thirty-seven 4-6-0Ts were acquired from Kitson and Stephenson, giving an appreciable advance in size and power. One of these, NGR No. 13 (Kitson 2269 of 1879), running as Escom No. 13 at Rosherville, near Johannesburg, recently celebrated its 100th birthday to become South Africa's only active centenarian locomotive – not bad for a heavy smoker.

From 1888 onwards the standard Natal main-line engine of the pre-Hendrie era appeared. This was the Dübs 'A' 4-8-2T designed by Milne, of which 100 were supplied between 1888 and 1900. One Dübs 'A' survived in service into 1980, at Springfield Colliery near Grootvlei.

In 1896 G.W. Reid took over from Milne. His first design was a stunning 4-10-2T, the first ten-coupled engine on the Cape gauge, which should have influenced locomotive design in South Africa far more than it did. Because it was such an advance on the Dübs 'A', management ordered only a prototype from Dübs in 1899, but this was so successful that another 100, known as the 'Reid tanks', were introduced between 1900 and 1903.

The CGR and NGR had evolved their own types from the early standard imported designs; and, while the CGR soon discarded tank engines for all but the lightest duties, Natal had only tanks until 1904.

This exclusively tank stud was not unique, for the NZASM used only tank engines throughout the 13 years of its existence.

Unlike the railways of the colonies, the NZASM ordered all its locomotives to factory designs – mostly by Emil Kessler. Five 0-4-0Ts, delivered for the 'Rand Tram' in 1889, were the first locomotives in the Transvaal. These were followed by a succession of small engines, culminating in the standard NZASM 0-6-4T, of which 191 were supplied from 1892 to 1899. Although reputed to be rough riders, many of these NZASM 'B's survived on the CFM and on some Transvaal gold mines until the mid-1970s. One is preserved in working order by the SAR.

For the rack section between Waterval Onder and Waterval Boven four 0-4-2T rack tanks for banking up the escarpment were supplied by Kessler between 1894-97. On the downgrade they were attached to the front of trains using their counter-pressure brakes for control.

In 1896, when the Orange Free State exercised its option to buy the CGR's main lines between the Orange River and the Transvaal border, 25 varied engines were bought second-hand from the CGR. These proved insuffícent and a further 34 6th class were delivered to the OVGS (Orange Vrijstaatse Goeverment Spoorwegen) by Sharp Stewart, Neilson Reid and Dübs in 1897-98. The last three arrived after the outbreak of war and were commandeered by the Imperial Military Railway (IMR).

The Anglo-Boer War broke out in 1899 and lasted for three years. Yet instead of the war there might have been another railway from Johannesburg, for at one stage Kruger agreed to a qualified franchise for 'uitlanders' (foreigners) if Britain allowed the Transvaal access through Swaziland to the Indian Ocean.

During the war the railways were vital in transporting troops, animals, munitions and supplies. The British immediately formed the 'Imperial Military Railways' and placed the CGR and NGR under military command, though their operation was left in civilian hands. Only the railways of the two republics came under direct British military control as the army advanced into Boer territory. Bridges, track and rolling stock were destroyed and any undamaged locomotives were worn out by wartime traffic and lack of maintenance. To provide motive power, the IMR urgently ordered 100 new engines – comprising Cape 7th and 8th class and Reid tanks – and even commandeered six Western Australian K class 2-8-4Ts en route to Freemantle.

After the Treaty of Vereeniging in 1902, when the Transvaal and Orange Free State came under colonial rule, their railways reverted to civilian control under a new state-run organization: the Central South African Railways (CSAR). The CGR and NGR reverted to their pre-war status.

For motive power the war marked a watershed, bridged only by Beatty of the CGR, and even his post-war designs were radically different. The period until Union in 1910 was a time for experiment, with many new designs most of which were highly successful and were produced extensively. The majority gave more than 50 years' service and most survived into the 1970s.

In 1901 Alco delivered the first of another successful series, the Cape 8th class – 16 bar-framed Consolidations. Beatty improved their riding by adding a four-wheeled leading bogie and shortening the driving wheel base. Sixty-one of these classic 8th class were delivered by Neilson Reid in 1902-03 and ten more were supplied by the newly-formed North British Locomotive Co. (NBL) in 1904. Eventually this builder became South Africa's largest supplier of locomotives, delivering more than 2 000 units before 1950.

Beatty decided the 8th class design could be further improved if he could increase the grate area. In 22 engines supplied by Kitson and NBL he reverted to the 2-8-0 arrangement and raised the boiler sufficiently for the firebox to spread out over the rear drivers, giving a 55% increase in grate area to 3 m² (32 ft²). The change in wheel arrangement allowed coupled wheel spacing to the Alco design of 1901, permitting an ashpan of greater volume. Nevertheless, the classic 8th class outlived the wide firebox variety by more than 40 years, and some were still in industrial service in 1980.

After the final batch of 8th class was delivered, Beatty discarded the narrow firebox and developed the wide firebox over trailing bissel wheels, which eventually became standard for more than 1 800 SAR locomotives.

Having seen the 'Japanese' Atlantics at work, he ordered the first wide firebox, bar-framed engines to his own design – the batch of seven 2-6-2s built by Neilson in 1901. Although classified as Cape 6th class, this design and later sixes were very different from the original and, presumably, were given the same designation as the classic 4-6-0s because they were intended for the same type of service. They were soon given another pair of carrying wheels under the firebox, creating the rare arrangement for a tender engine of 2-6-4. It was with this design that Beatty introduced the so-called 'bridle casting', used in new designs for more than two decades. In the last two 6th class variations, involving 8 engines, Beatty experimented with larger driving wheels and used piston valves and superheating for the first time in South Africa; he also introduced the Pacific wheel arrangement, known as the 'Karoo' type.

Three more experimental designs decided him on the ultimate CGR main-line power. His big 2-8-2 and 4-8-0 of 1906 were both supplied by Kitsons with wide fireboxes, one deep, the other shallow. Extensive testing finally convinced Beatty that the deep, wide firebox suited South African coal, though it was the class 10, with the shallow firebox, which was adopted as a standard design in both Rhodesia and Angola. The third experiment was a 3-cylinder

289

290

291 292

293

294

291. Tank engines have featured prominently in South African locomotive history and at the turn of the century two railways, the NZASM and the NGR, were operated exclusively by such engines. The most numerous single class was the '46-tonner' – the ZASM denoting their engines by mass – of which 191 were built. Today one has been preserved in working order and she is often used for filming, as was the case at Merhof, Transvaal, in June 1979.

292. Before its first tender locomotives, the NGR's heavy main-line locomotives were ten-coupled machines, the 'Reid tanks'. These 101 engines, though massive for their time, were quickly outclassed with the arrival of larger, though eight-coupled, tender locomotives. In later years most of these 'Reids' were converted to 4-8-2Ts and sold to various mines; however, several lasted in SAR service at the coal terminal on the Bluff at Durban, until the mid-1970s.

293. Hendrie's standard eight-coupled main-line locomotives of the pre-1920 period were all cast in a similar mould, with plate frames, low running boards, and Belpaire boilers. No. 1970, the last Class 15A was still in service in 1980, used for shunting at De Aar and also available for steam specials.

294. From the early 1930s to the mid-1950s more than 480 Hendrie engines were reboilered and fitted with new 'standard' cabs which very much changed their outward appearance. No. 1911 is a 14R stationed at De Aar, and uniquely fitted with a torpedo tank from a Class 24 2-8-4.

295. The 6th Class 4-6-0 is representative of the best in pre-20th century Cape Government steam development. Though displaced from the most important services by 1910, it lasted well into the 1950s on suburban work, particularly around Cape Town, Port Elizabeth and Johannesburg. Their final work was in shunting at Cape Town and Port Elizabeth and even though SAR's last examples were retired in 1973, one engine still survived on a Transvaal colliery in 1980. Some acquired Belpaire boilers designed by P.A. Hyde.

296. The first Twentieth Century CGR passenger design – the SAR 5B, or 'Karoo' Pacific – was very advanced for its time . . . and a taste of things to come. No. 523, built by Beyer Peacock in 1904, is on display, preserved at De Aar.

297. The ultimate Pacific type – the 16E. Wide firebox, generous boiler, bar frames, high wheels and RC poppet valve gear all characterize this magnificent design. The one surviving operational locomotive is still capable of speeds from 120 to 140 km/h with heavy loads.

298. Built for power – and displaying its brute strength: the GLs are usually compared to EAR's 59 Class, but come off quite well even in comparison to the solitary Russian Garratt – the largest of these articulateds ever built.

295

296

297

298

Smith compound using the Karoo layout. The fuel savings did not match the additional cost of maintenance and this design was not repeated.

These three experiments gave Beatty all the information he needed to put the final CGR classes on the drawing board – a heavy 4-8-2 for mixed traffic and an enlarged Karoo type for passenger service. He completed the designs early in 1910, but both new types were delivered after he had retired in May 1910, soon after the formation of the SAR. The two 4-8-2s arrived from NBL in 1911, and the 4-6-2s, of which four were built, from Vulcan in 1912.

The last design directly attributable to Beatty was SAR class 4A, which arrived in 1914, four years after his retirement. Ten were delivered by NBL and they were an updated class 4, with superheaters, piston valves and Walschaerts valve gear. The entire class of these long-lived engines survived into the 1970s. They formed the basis for the Rhodesian Railways 10th class.

When the CSAR took over from the IMR in 1902 and P.A. Hyde was appointed chief locomotive superintendent, he ordered 60 Cape 8th class. He also designed an excellent Belpaire boiler of increased capacity for his inherited 6th class engines, so that the reboilered sixes could be used on the same loads as the 8th class locomotives.

With his class 9 Pacifics and F class 4-6-4Ts of 1904, Hyde began quietly to demonstrate his competence. The five Pacifics were unique to South Africa in having narrow fireboxes. Nevertheless, with the good Witbank coal of those days they performed well for a quarter of a century. The F class tanks – known as 'chocolate boxes' because of their immaculately lined black finish, with polished brass dome and copper-capped chimneys – were used on the Reef suburban services.

His next two designs were altogether different. Thoroughly modern in concept, they stood up to comparison with the finest contemporary European and American practice. Class 10 was an express passenger Pacific designed to take advantage of the 40 kg (80 lb) rail with which the CSAR was then relaying their main lines. These Pacifics had large-diameter, long-travel piston valves which gave an extraordinarily free-running machine. Fifteen were delivered by NBL in 1904 and all put in more than 50 years of service.

Class 11, also built by NBL in 1904, was the CSAR's heavy freight engine. With the same modern cylinder design as the Pacifics, these Mikados were also excellent runners, in spite of their 1 220 mm (4′ 0″) diameter driving wheels. A pair of 11s on a freight train were paced at 100 km/h in 1973, when they were in their 70th year. In CSAR service they were allowed 950 tonnes on 1-in-100 gradients, barely within their tractive effort (as built) of 16 685 kg (36 800 lb). Of the original 36 engines, 35 survived into the 1970s and many are still used on mines today.

G.G. Elliot who followed Hyde, understudied H.M. Beatty on the CGR and brought with him to the CSAR some of the great man's ideas.

Elliot was probably the first in South Africa to realize the importance of long valve travel for expansive (and therefore economical) steam working; but his main contribution was to introduce superheating as a matter of policy. In 1909 Elliot designed an updated version of Hyde's class 10. Ten were ordered from NBL, of which five were equipped with Schmidt superheaters. According to the CSAR General Manager's Report of 1910, those 'fitted with Schmidt's superheater . . . may be taken as being 25 per cent better than similar engines not so fitted. The increased load hauled by those engines has almost entirely dispensed with the necessity for double-heading passenger trains in the Orange Free State.' After this, superheating was introduced on the more modern existing classes and, with two exceptions, on all new classes. A further five of the superheated Pacifics were delivered in 1912.

On an American visit, Elliot placed orders with Alco for two engines of that company's design. The first was a bar-framed Pacific based on class 10, but with a much bigger boiler and heavier axle load. It was the first passenger engine in South Africa with a tractive effort exceeding 13 600 kg (30 000 lbs). The second was a 2-6-6-2 Mallet, in many ways the most impressive locomotive in South Africa until then. It was put to work on the increasingly busy Witbank coal traffic in 1910. The ruling grade facing loaded coal trains between Witbank and Germiston is an 8 km stretch of 1-in-100 compensated and the Mallet, with its rated tractive effort of 50 448 lbs (22 900 kg) soon showed that it could muscle 1 600 tonnes over this road, as compared with 950 tonnes for an 11th class. Repeat orders were placed with Alco for 14 superheated engines, classified SAR MF. The prototype unsuperheated machine became class MD.

The final Elliot design to enter service was the class 10C light Pacific, of which 12 were delivered by NBL in 1910 and 1911. They were intended for Reef suburban services but were so successful that they put in many years of main-line work on fast passenger trains. They were remarkably rapid and free running for an engine with only 1 450 mm (4′ 9″) driving wheels, a characteristic again attributable to Elliot's excellent cylinder design.

On the NGR Reid was succeeded by D.A. Hendrie who, in a country noted for its distinguished mechanical engineers, towered above them all. If longevity is a criterion, Hendrie's designs, prepared entirely within the first two decades of the century, may be among the finest ever built. Most survived into the 1970s and large numbers into the 1980s – examples of nearly every class he designed are still in use either on the SAR or in industry.

When Hendrie took office the increasing demand for coal at Durban, both for bunkering and export, was creating chaos on the Natal main line. Coal trains needed a 'Reid' banked by two Dübs 'A's on the 1-in-30 uncompensated up from Estcourt to Highlands. Hendrie broke Natal's 44-year tradition of tank engines and introduced the Hendrie 'B' 4-8-0, later to become SAR class 1. Fifty were delivered by NBL in 1904, greatly improving the flow of traffic. After two years, six engines were equipped with trailing bissels behind the firebox to steady the engines when working passenger trains, becoming the earliest 4-8-2, or Mountain type – a wheel arrangement which eventually graced more than 1 400 SAR locomotives. A further 21 Hendrie 'B's were ordered from NBL in 1909, the last being fitted with piston valves.

The Mountain type was eventually to become as much a part of the Natal scene as bananas. Hendrie's 'D' class 4-8-2s were introduced between 1909 and 1910 when NBL supplied 30. They had slide valves and were initially unsuperheated.

Hendrie's 1909 visit to the USA resulted in an order for two locomotives, a superheated 4-8-2, based on the Hendrie 'D', and a Mallet for main-line banking out of Estcourt. This 2-6-6-0 was much smaller than the CSAR's Mallet and was unsuperheated. These were the last placed in service on the independent NGR.

When on May 31, 1910, the Union of South Africa was forged, the CGR, NGR and CSAR merged to form the South African Railways (SAR), with Hendrie as chief mechanical engineer.

The SAR came into being with a total of 1 460 locomotives of more than 110 different classes. Of these, 255 engines were declared obsolete, though most continued to be used for shunting for some time after Union.

The twelve years of Hendrie's term as CME saw the annual freight tonnage of the SAR move from 12 000 000 to 20 000 000 tonnes and passenger journeys from 33 million to 64 million; to keep pace with the increasing traffic the locomotive stock increased to 1 735 in 1922 and the average tractive effort rose to 13 600 kg (30 000 lbs).

Hendrie developed the Mallet design for several years, until the arrival of the first main-line Garratt in 1921 demonstrated the superiority of this form of articulation. Fifteen 2-6-6-0s, all saturated, were supplied by Alco and NBL in 1910 and 1912.

Hendrie now decided to meet the challenge of a statement by the general manager in 1911, that 'engines of 60 000 lbs tractive effort have been offered by British and American builders', by designing such an engine himself. He set up a team with E.H. Mellors and J.R. Boyer, the chief locomotive draughtsman and his assistant, and they worked many hours into the nights of 1914 preparing detailed drawings for what was at the time the heaviest and most powerful locomotive built outside of North America.

NBL supplied five of these MHs in 1915 and they were placed in service on the Witbank-Germiston coal line. Their axe-load of 18,5 tonnes was not exceeded on the SAR until the arrival 12 years later of class 18. Their low pressure cylinders of 800 mm (31½″) diameter were the largest ever applied to a narrow-gauge locomotive. After a few years they were transferred to the Vryheid-Glencoe coal line in Natal, where their boiler pressure was reduced and the maximum starting tractive effort was correspondingly reduced to 24 300 kg. In this service they successsfully handled 800 tonnes on 1-in-50 uncompensated gradients for 16 years, until ousted by the GLs in 1937-38. Until 1918 another 41 Mallets of various types were supplied for main-line banking and branch-line services, the last being scrapped in 1962.

The first SAR Garratts went into service in 1919. They were three 2-6-0+

0-6-2s for the 600 mm lines and were followed by an ultra-light 2-6-2+2-6-2 and a heavy 2-6-0+0-6-2 for 1 067 mm-gauge lines. The 600 mm engines, classified NGG11, were used on the Umzinto-Donnybrook line, which climbs from sea level to 1 366 m in 150 km. The light Garratt became class GB and was put into South Coast traffic; the heavy one became class GA. In comparative trials with an MH it was so successful that no more Mallets were ordered, though 84 were already in service.

Of the assorted engines which Hendrie inherited, the passenger types were probably the best and not until 1914 was new passenger power needed. Fifty-four Pacifics of the various classes 16 were supplied by NBL between 1914 and 1921, the last batch of 30 having combustion-chambered boilers. Class 16A's two locomotives were of special interest in having four cylinders – rare for a 1 067 mm-gauge locomotive. Hendrie's Pacifics were all successful, only being removed from main-line passenger service from the 1950s when the gradual introduction of steel coaches created trains with masses beyond their capacity. All worked into the 1970s, except for the two four-cylinder machines and one other, destroyed in a collision.

Of all Hendrie's successful designs, the series of Mountain types which he conceived between 1909 and 1919 were among the finest. With the 'D' class of the NGR he had arrived at his standard heavy freight and mixed traffic wheel arrangement. His new 4-8-2s fell into four basic categories, each adapted to the geographic conditions in which they were to operate.

Classes 3, 3B, 14 and 14B – 100 engines in all – were designed specifically for the Natal main line, with its tough parameters found nowhere else on the SAR – 100 m-radius reverse curves on 1-in-30 uncompensated gradients, with permissible axle-load of 16,5 tonnes. Class 14A was designed for the 30 kg track of the Cape Eastern main line; it was basically the same below the waist as class 14, but had a smaller boiler and two tonnes less axle-load.

Classes 12 and 12A, comprising 113 engines, were designed for heavy haulage on the Witbank-Germiston coal line. In service, the 12s managed only 1 270 tonnes, compared with 1 600 tonnes for an MF, but they did the 260 km round trip in one shift, which the Mallets could not do. When the 12As arrived in 1919 they could handle 1 450 tonnes eventually putting paid to the Mallets in coal-line service. Class 12B, of which 30 were supplied by Baldwin in 1922, was an updated 12, designed for the Midland main line, which did not have the severe curvature of Natal.

Perhaps the finest Hendrie 4-8-2s were his mixed traffic class 15 and 15A; these 129 engines were intended for the Cape main line south of Kimberley and the Free State main line south of Bloemfontein.

Altogether 413 SAR 4-8-2s were built to Hendrie's designs and a further 103 to his specifications. Of these 516 Mountains, 510 remained in daily service on the SAR into the 1970s, another was in industrial use. In more than 60 years, only five Hendrie 4-8-2s were withdrawn, and then only because of accident damage. As the 80s dawn, more than 400 Hendrie 4-8-2s are still in service, in spite of the previous decade's massive dieselization which accounted for entire blocks of subsequent classes.

Hendrie retired in 1922 and was succeeded by Col. F.R. Collins who drastically changed CME policy. Whereas Hendrie had prepared designs to the last detail, Collins preferred merely to draw up the bare specifications and let the builder carry on with the detailed design work. This prevented standardization, and large numbers of new classes proliferated.

The Collins era marked the first major switch from British to Continental and American builders, further aggravating the problem of new types. Between 1922 and 1928 a total of 25 new classes came onto the SAR roster.

Branch lines in Natal needed a machine with a light axle-load and Collins allowed three variations, all with the 2-6-2+2-6-2 arrangement, to be produced. The first was a classic Garratt of the period, with Belpaire boiler, plate frames and other tried and trusted features of Gorton. Six of these GCs were supplied in 1924. At the same time NBL offered an engine with identical boiler and engine unit dimensions, but articulated on the Kitson Meyer principle. One was ordered for comparison with the Garratt and, since the makers had coined the name 'Modified Fairlie' for the type, it was classified FC. Three years later Krupp supplied 39 class GCA engines with almost identical dimensions as the GC, but with bar frames and round-top fireboxes.

The GCAs, designed for 22,5 kg track, were very successful, and the entire class gave more than 40 years' service until the elimination of 22,5 kg track and dieselization led to their withdrawal.

299. **South Africa's largest Mallet, Class MH. Note the huge (800 mm-diameter) low pressure cylinders.**

For heavy freight haulage on 30 kg (60 lb) track, the GE 2-8-2+2-8-2 was introduced in 1925. In the period up to 1930 Beyer Peacock supplied 18, each batch having minor modifications. It also appeared in two other guises. The first was the class HF (for Henschel Fairlie), of which 11 were delivered in 1927. Their engine units were virtually the same as the Garratt's, but inexplicably the boilers were much smaller – another example of the confusion and lack of direction in SAR locomotive policy at this time. The final GE variation came only after the war, as class GEA. Like the GCA it had bar frames and round-top boiler.

For certain lines the chief civil engineer had specified an immediate axle-load between the 10 tonnes of the GC and the 13 tonnes of the GE. The result was the GD, 14 of which were built by Beyer Peacock in 1925. Again, two variations derived from this type. NBL supplied four of their 'Modified Fairlie' version in 1926, classified FD, and in 1929 Linke Hoffman delivered five of a bar-framed round-top-boilered version, class GDA.

In 1924 Collins sent one of his locomotive superintendents to the USA with plans and specifications for heavy main-line mixed-traffic and passenger locomotives. This led to orders for two 4-8-2s and two 4-6-2s being placed with BLW to their detailed designs. The Mountains became class 15CB, and the Pacifics 16D. This was the first big American power seen in South Africa, and incorporated many features common in America but not previously seen in South Africa – arch tubes, master-mechanic self-cleaning smokeboxes and grease lubrication; the 4-8-2s also had combustion chambers. Ten more 15CBs and five 16Ds were delivered by BLW in 1926.

The 16Ds worked the 1 536 km between Cape Town and Johannesburg, with only one engine change – at Beaufort West; the 15CBs worked the 545 km stretch between Beaufort West and Cape Town, which included the Hex River Pass and the Pietermeintjies bank.

The 15CBs and 16Ds marked a watershed in SAR design, making the final break from the plate frames and low boiler centre-line of Hendrie. All future main-line power had high-pitched boilers with bar frames.

The 12 15CBs performed so well that by 1930 orders for a further 84, with improved frame design, had been placed with BLW, Alco, Breda and NBL; they were classified 15CA. In their final form all 96 engines had 5′ 0″ (1 524 mm) driving wheels and a tractive effort of 21 800 kg (48 100 lbs). After more than 50 years the 15CBs were withdrawn from SAR service, many being snapped up by industrial railways; all but three of the 15CAs were still in SAR service in January 1980. Similarly the success of the 16Ds led to repeat orders for 14 engines from Hohenzollern and Baldwin in 1928. Only a few survive – on industrial railways in South Africa and Zimbabwe.

In 1927, two large three-cylinder 2-10-2s of class 18 were delivered by Henschel for the Witbank coal traffic. Their starting tractive effort of 27 200 kg (60 100 lbs) made them the most powerful non-articulated locomotives ever to run in the southern hemisphere. In service their Gresley gear was found to be heavy on maintenance and the design was never repeated. Nevertheless, the two gave more than 20 years of service.

What was to be the largest Garratt class anywhere for more than 20 years made its debut in 1927, with the delivery of 65 class GFs from Hanomag,

300. **Largest of the 'modified Fairlies' were these HFs by Henschel.**

301. **Maffei built these 'U-boats' in 1927. A combination of the Garratt and modified Fairlie principle, they were never outstandingly successful.**

Henschel and Maffei. With the exception of two engines scrapped after accidents and four sold to the Caminhos de Ferro de Moçambique after the war, they all worked for the SAR for over 45 years. Many survive in industrial use.

The year 1927 saw yet another Garratt variation, with the introduction of class U, a 2-6-2+2-6-2 'Union Garratt'. In this combination of the Garratt and Kitson-Meyer principles, the front unit pivoted with its tank while the back unit, as on the modified Fairlie, pivoted under the main boiler frame extended backwards to carry the coal bunker and rear water-tank.

The GHs, from Maffei in 1928, were two more engines on the 'Union Garratt' principle. They were large-wheeled 4-6-2+2-6-4 passenger machines. With their 18-tonne axle-load and a 22 800 kg (50 400 lbs) tractive effort they were the most powerful South African passenger locomotives. Like the Us, the GHs survived into the 1950s.

An important new family of branch-line engines was founded in 1928 when four light 4-8-2s of class 19 by Schwartzkopff arrived. They were an immediate success and in 1929 SLM followed up with 36 of a slightly lighter version, with smaller boiler and wheels. A further 14 class 19s, slightly modified, were delivered in 1930 and classified 19B.

Of the 25 different classes introduced under Collins, not all were successful – and he was mostly to blame. In 1928 he partly redeemed himself by ordering, to his specifications and Beyer Peacock's design, the magnificent class GL. With 40 375 kg (89 130 lb) of tractive effort, they share the laurels with EAR class 59 as the most powerful engines ever to run south of the equator.

While the 59 class had 3% more heating surface, it had 4% less grate area; and though it had 10% more adhesion mass, it had 7% less tractive effort. GLs rated 1 132 tonnes on 1-in-50, while 59s rated 1 216 tonnes on 1-in-60. GL boilers were more than twice the rail gauge in diameter and, with 7,0 m^2 (75 ft^2) of grate, were the largest ever applied to a South African locomotive. Eight were built by Beyer Peacock in 1929-30 and performed superbly between Durban and Pietermaritzburg. The line from Pietermaritzburg to Ladysmith had been electrified in 1925 and when electrification was extended to Durban in 1937, the GLs were transferred to the Glencoe-Vryheid line. Here for 30 years single engines coped with 1 132 tonne coal trains on gradients which were 1-in-50 uncompensated, until eased on an almost entirely new location prior to electrification. The class was retired in 1972, but two have been preserved.

In 1929 the first locomotives in South Africa designed solely for shunting arrived in the shape of 14 Henschel 0-8-0s. As delivered, they had a hefty 23 300 kg (51 400 lb) tractive effort, but this was reduced to 14 400 kg in 1933. The S class spent its entire SAR life on the Witwatersrand and, when retired after nearly 50 years, some were taken on in industrial service.

In 1929 A.G. Watson took over from Collins. Under his direction the motive power policy was kept on a tight rein, assisted by the Depression of 1930-33. Only six entirely new classes were added during his era, but a range of 'new' types resulted from Watson's policy of standardization of locomotive boilers. Lovers of Hendrie's beautiful designs deplore this, but at the time it made economic sense.

Other Watson legacies were his large grate areas and high-pitched, large-diameter, skyline-filling boilers. By and large these were successful, and some fine-looking modern designs resulted from these progressive features. His decided ideas on how to burn South African coal were embodied in the first locomotives into service after he took office. The grate area of the six Henschel 16DAs was increased by 20% to 60 ft^2 (5,6 m^2) – unfortunately for the fireman they were hand fired, but they steamed well. The 16DAs were withdrawn during 1973, but No. 879 is used for enthusiast specials.

In the early 1930s many railways experimented with poppet valves. Some, notably the French, had achieved good results and, when a further batch of 50 branch-line 4-8-2s was needed, Watson instructed NBL to equip them with rotary-cam-operated poppet valves, at an additional cost of £200 for each engine. The result was the eminently successful class 19C. For more than 40 years, mainly on the long branches out of Cape Town, these free-running and speedy engines repaid with interest the extra money spent on them.

In 1938 they were indicated during comparative steam consumption tests with the Walschaerts-motioned version, class 19D, and some diagrams were obtained showing extraordinarily low cylinder efficiency. In view of the consistently high performance of the r.c. engines, we are tempted to assume that the officer who indicated the test engine did not know what he was doing. All except one were withdrawn in 1977; today No. 2439 is kept in sparkling condition for working special tourist trains.

After the Depression, the railways of Europe and America turned to speed to combat increasing competition from road transport. Not to be outdone, the SAR management asked the chief civil engineer to see if a 70 mph (112,5 km/h) limit could be introduced on certain main lines. An ambitious map showed long sections of the Cape and Free State main lines and the Natal main line as far as Volksrust which, it was considered, could be brought up to the standard required. At the same time Watson was instructed to design an express passenger engine. The majestic 16E was the result, six being supplied by Henschel in 1935.

Unfortunately, the programme to increase the permissible main-line speeds was not implemented until after the 16Es retired in 1973. Then the speed was raised to 110 km/h on certain sections for the Blue Train only. Nevertheless, in spite of the official limit of 90 km/h, there are many recorded runs of more than 110 km/h by 16Es – during their first trials and later. Two of these remarkable engines are preserved today. One (No. 858) in working order, is regularly in demand on enthusiast specials.

The freight/mixed-traffic class 15Es soon followed their glamorous sisters, 44 being delivered by Stephenson, Henschel and Berliner Maschinenbau in 1935. They spent the first 20 years of their lives in the Karoo, and the next 20 in the eastern Free State. No. 2878 is preserved in running order.

Watson made the tractive power of a 15E available for branch lines when he designed the class 21, a solitary 2-10-4, for 30 kg rails. Although promising, the design was not repeated because Watson retired in 1935, before its introduction, and his successor, W.A.J. Day, returned to the articulated power which Watson had opposed.

Day disliked poppet valves and his period in office saw a reversion to piston valves and Walschaerts motion. He argued that large diameter long-lap, long-travel piston valves could perform as efficiently as the NBL-designed poppet valves – and this was borne out in 1938 when tests on the first batch of class 19D showed a slightly lower specific steam consumption than class 19C.

The 19D became the most common branch-line locomotive on the SAR. The first were delivered by Krupp in 1937 and after batches by Borsig, Skoda and Stephenson, the last of 235 engines arrived from NBL in 1948. They have worked throughout South Africa and still survive almost intact though since 1972 they have been hounded continually by class 35 branch-line diesels.

A new Garratt design was introduced in 1938. The GM was ordered for the Krugersdorp-Mafeking run in preference to the class 21, as Beyer Peacock had indicated that they could provide an engine with the power of two 19Ds. Sixteen of these performed usefully for more than 35 years.

While the 19D is the piston-valve version of the 19C, the 15F is the piston-valve version of the 15E. In 1938 Berliner Maschinenbau built the first 15Fs – the most numerous SAR class, which eventually totalled 255 engines, supplied in several batches by Henschel, Beyer Peacock and NBL until 1948. All but the first 22 were equipped with mechanical stokers. Big brothers to the 15F were class 23, of which 136, all mechanically-fired, were supplied immediately before World War II, proving another successful design. All 23s and all the 15Fs were intact in January 1970, but by January 1980 only one 23 remained in service, as compared with 250 15Fs.

Just prior to World War II, Day was succeeded by M.M. Loubser, father of the present general manager. At first Loubser contented himself with detail improvements, initially to the existing 19D and 15F designs for repeat orders, then to a new standard boiler for Hendrie's 12A. All 12ARs were still in use in 1980.

Towards the end of the war, plans were drawn for an updated GE with 4-8-2+2-8-4 wheel arrangement, bar frames and round-top firebox. Fifty of these GEAs, from Beyer Peacock in 1946, gave 30 years' service in Natal and the southern Cape before their withdrawal.

The practice of using down-graded main-line power for shunting was carried to extremes on the SAR, and when some of the pensioners used for shunting were getting too long in the tooth, Loubser designed the S1, a heavy 0-8-0 with enormous boiler, based on his standard type for class 12A. Twelve were built at the Salt River works in 1947 and, apart from the mass production of the standard Watson boilers, this was the biggest job undertaken in SAR workshops up to that time. A further 25 of these powerful engines were supplied by NBL in 1948 and all S1s were still in service in 1980.

Though the 19th classes had almost solved the problem of traffic on secondary lines, they were restricted to 30 kg rail and thus barred from the many kilometres laid with 22 kg and lighter rail, including the whole of South West Africa, still worked by the aging classes 6 and 7. Accordingly, a lightweight 2-8-4 – with cast steel beds being specified for the first time on a South African locomotive – was designed, resulting in the fine class 24, of which NBL supplied 100. No. 3675 of this batch was the 2 000th locomotive supplied by this firm to the SAR. Dieselization and rapid replacement of 22 kg rails made many 24s redundant in the late 1970s, but several were still at work in 1980.

The war and post-war years saw unprecedented traffic increases, a large proportion of the extra burden being carried on the Cape main line. Here the powerful 23s and 15Es coped well, but water was becoming important in the motive power-traffic equation, for 700 km of the Cape main line runs through the southernmost and driest region of the Great Karoo. Water supplies, almost entirely from boreholes between Touws River and De Aar had sunk dangerously low and, with traffic approaching an annual 20 million tonnes, it was decided to introduce condensing locomotives on this stretch.

In 1948, just before Loubser retired, he arranged for Henschel to supply a condensing tender and artificial draughting equipment to convert Watson's solitary class 20, built in 1935 at Pretoria workshops as an experimental light axle-load 2-10-2 for service in South West Africa. It utilized the modified 19C poppet-valved cylinders and wheels of a scrapped 8th class and by the time the Henschel patent condensing system was added, the first SAR condensor was rather a mongrel, but it was the progenitor of a fine strain. When it went into service in 1950 it could easily cover the 550 km between Touws River and De Aar on one tenderful of water. By 1951 it had supplied sufficient data to enable Henschel to proceed with the design of class 25 condensing 4-8-4s.

Much groundwork for the basic design had already been done by SAR engineers, notably H.J.L. du Toit, who had developed the proportions of the huge 25NC boiler from tests conducted in the late 1940s with a class 23 experimentally equipped with a combustion chamber. Roller-bearing specifications came from the USA and virtually every moving part on the 25s – from the main driving axleboxes to the little-end gudgeon pins – is equipped accordingly. Other welcome American features were the cast-steel mainframe incorporating the cylinders, mechanical lubrication and self-adjusting wedges.

By the time orders were placed for the 4-8-4s, L.C. Grubb had succeeded Loubser as CME.

A non-condensing version of the 25, classified 25NC, was ordered at the same time, and the total order of 140 engines (90 class 25s and 50 class 25NCs) was shared by Henschel and NBL.

The first condensing 4-8-4, No. 3451, was assembled at Salt River and went on its trial run in 1953. It made an inauspicious start – the valve gear seized up in the first 20 km. This was the prelude to several years of teething troubles.

The root of the problem lay in the size of the new engines. The 25 class boiler could generate more than twice as much steam as the experimental class 20 and nearly double the capacity of the Kriegslok. Up to 25 000 kg of steam an hour had to be converted back to water, in ambient temperatures as high as 40°C. For this a condensing tender was provided to house five huge cooling fans, 2,1 m in diameter, driven by an exhaust steam turbine, 16 banks of condensing elements (eight each side), and condensate tanks. These, with the coal supply and the mechanical stoker, required a tender 3 m longer than the engine. The tender gave few problems, but the locomotive developed several which only came to light when they went into regular service.

One problem that was soon sorted out was the accumulation of char in the smokebox. The original smokebox was extended by 300 mm with a V-shaped receiver for the cinders underneath, creating the ugly banjo-front characteristic of these engines.

Exhaust steam on its way to the tender drives a turbine which in turn is coupled to the draught-inducer fan. The speed of rotation is thus directly proportional to the flow of exhaust steam and therefore the draught through the firebed is controlled as in a normal locomotive.

Initially, the fans wore out rapidly and unevenly because char cut the blades. Worse still, this wear caused imbalanced stress on the bearings, which frequently failed as a result. This problem was largely alleviated by altering the blade angles so that fan replacements were needed only every 50 000 km.

The third main difficulty encountered was cylinder and valve lubricant in the condensate. It was thought that if too much oil was in the feedwater it would adhere to the firebox, causing crownsheet collapse and consequent boiler explosion because of the insulating effect of the oil.

What with water and mechanical problems, engines managed no more than a few hundred kilometres a month, and there is an immediate parallel with the story of the Algerian passenger Garratt. Both were powerful locomotives of innovative concept made virtually ineffective by teething troubles and the pampering they needed. Fortunately the parallel ended there, for engineers and chemists persevered unremittingly until they solved the oil problem and justified the original decision to use condensers in the Karoo.

After 1955 the Karoo water stations were always able to meet the demand and never again was it necessary to run water trains, a practice which had become common during the war. It was calculated that even in the hottest weather a 25 could save 85% of its water and there was a bonus of 10% coal saving owing to the use of heated feedwater.

By 1954 the 25NCs had all entered service and were an unqualified success. They did the 700-km run from De Aar to Welverdiend (near Johannesburg) unchanged, recoaling on their trains at Warrenton; moreover, at Welverdiend they were merely turned, coaled and watered before being sent south again on the first available train. Their annual kilometrage has always been dictated by utilization rather than availability.

Steadily, the field of operation of the 4-8-4s has been shrunk, first by electrification and later by dieselization. When diesels took over between Beaufort West and De Aar in 1973-74, it was decided to remove the condensing feature from the 25s bringing them into line with the NCs which needed less maintenance. By January 1980 all but three had been converted – two are to be preserved and one will remain working as a condenser on tourist trains.

Meanwhile the future of the 137 class 25NCs has brightened for they are now more economical than diesels and their failure in traffic rate is half that of five-year-old diesels.

The introduction of the gas-producer combustion system on these engines will make them cheaper than both diesel and electric traction. The conversion of the first engine to the designs of Wardale has already begun.

The years 1953-54 saw the introduction of three new classes – S2, GMA and GO. Krupps built the 100 light 0-8-0s, classified S2, for dockside shunting; they were intended to replace the armies of classes 6 and 7 still used for such work, but the increased mass of the trains and the introduction of diesel shunters reduced their usefulness and only a few remained at work in 1980.

The class GMA and its heavier axle-loaded counterpart, the GMAM, became the most numerous Garratt type, 120 being built by Beyer Peacock, NBL and Henschel between 1954 and 1958. These were the last 1 067 mm-gauge steam locomotives supplied to the SAR to date. An excellent design, they have constantly been at a disadvantage in main line service because of their low axle-load. Nevertheless, on the East London main line they could handle 815 tonnes on continuous 1-in-50 compensated grades, at a constant speed of 18 km/h, and they were supremely effective on Montagu Pass, where they were rated at 500 tonnes on the mist-prone 1-in-36 grades. Apart from those destroyed in accidents, 111 GMAMs are still on the SAR roster, though with the massive dieselization of the 1970s most are in storage, but 22 were on load to Zimbabwe in 1980.

302. **The last of a noble strain. The very last Garratts built anywhere were the eight 600 mm-gauge examples built by Hunslet Taylor in Johannesburg in 1968. They are also the last new steam locomotives, to date, on the African continent.**

303. **Montagu Pass, the climb of the Outeniqua Mountains, was one of South Africa's most spectacular mountain sections. Garratts were the standard power for well over 50 years, with first GDs, then GEAs and finally GMAs being used. In January 1979, a GMA roars into tunnel 6. Almost one year to the day the last regular steam operation over the pass ended.**

The last new class on the SAR was class GO, the Garratt of light axle-load, of which 25 were built by Henschel in 1954. Having spent most of their careers involved with ore trains on the Steelpoort branch until displaced by diesels in 1972, they ended their active days in Natal in 1976, when they had barely been run in. All are now in storage.

Sub-Cape-gauge railways: The narrow narrow-gauge in South Africa

Although the main-line and trunk networks were well on their way to completion by the late 1890s, several important farming regions in South Africa continued to be served only by oxwagon transport. The reason for this was the cost of new construction on 1 067 mm gauge, which had escalated to more than £10 000 a mile. In his *Light Railways,* published in 1896, J.C. Mackay contended that a 600 mm-gauge line in the Long Kloof would be an economic proposition, with estimated annual receipts of £250 a mile, against an estimated overall construction cost of £3 000 a mile. A 1 067 mm railway was out of the question, and in 1898 the Cape Parliament authorized its first narrow narrow-gauge railway.

Long Kloof construction began only in 1902, by which time another 600 mm line, the Hopefield Railway – authorized only in 1900 – was well on its way to completion. Later construction in the Cape continued intermittently until 1927 when the Upington-Kakamas line was opened. At peak, the Cape Province had 580 km of state-owned and 60 km of private 600 mm-gauge lines and the Cape Copper Company's 175 km of 760 mm gauge between Port Nolloth and O'Kiep. Today only about 340 km of the 600 mm lines remain open – the thriving Long Kloof network, based on Port Elizabeth and running inland through the orchards to Avontuur, and the Eastern Province Cement Co.'s 25 km of private track at Port Elizabeth.

Natal followed the Cape's lead and by 1917 had 393 km of 600 mm gauge in operation, as well as about 500 km of more or less permanent track owned by various sugar mills. All the SAR and a considerable length of private 600 mm-gauge track remain in use.

The Orange Free State did not have sub-1 067 mm-gauge lines; the Transvaal had 100 km of state-run 600 mm gauge, but this has disappeared, as has almost all privately-owned 760 mm and 600 mm trackage.

The first locomotives ordered for the CGR's 600 mm lines were of two remarkably contrasting designs – three perfectly American-looking Moguls by Baldwin and two properly British 2-6-4Ts by Manning Wardle. Standard CGR narrow-gauge power until the arrival of the Garratts was various Bagnall, Kerr Stuart and Baldwin 4-6-0s, of which 15 were in service by 1915. The motive power shortage during World War I led to the acquisition of a batch of nine Falcon 4-4-0s from the Beira Railway; named 'Lawleys', after the engineer who built the Beira line, they had been in storage for 15 years. Six neat Baldwin Pacifics, which arrived in 1916, were to work the Avontuur line for more than 40 years, except for one which found its way to South West Africa.

Natal started operations on its 600 mm gauge with Hendrie 4-6-2Ts, which became standard power until the arrival of Garratts.

The NGG11 design of 1919 was so successful that Garratts became standard on all SAR narrow-gauge lines outside South West Africa. The first NGG13s arrived from Hanomag in 1927 and were followed by several repeat orders for the almost identical NGG16. When the South West African lines were widened in 1960-61, all their class NG15 2-8-2s were transferred to Port Elizabeth to work the Avontuur line, and were promptly nicknamed the 'Kalaharis' by the shed staff at Humewood Road, in spite of the fact that they had always worked near the Namib and not the Kalahari desert. The class NGG16s were bought from the Tsumeb Corporation, and these brought the total number of NGG13/16 engines in the Union to 36. In 1968 eight more NGG16s were built in South Africa by Hunselt Taylor, thus this remarkable Garratt design was produced for more than 40 years. These were the last new steam locomotives to be placed in service in Africa. Altogether some 144 engines of 30 classes worked the 600 mm-gauge lines of the SAR, with about 220 more engines on privately-owned 600 mm and 760 mm-gauge lines.

304. **Tunnel 5 was the place to be during mid-summer, when the Mossel Bay-Johannesburg train ran five days a week. Soon after 18h30 it burst out of the tunnel, with strong rays of afternoon light creating a memorable sight.**
305. **After the arrival of the larger Garratts, banking on the pass was unusual but in April 1979, when the Railway Society of Southern Africa ran a special 11-day steam excursion, the 'Sunset Limited' required the assistance of a 24 Class 2-8-4.**
306. **The cab interior of a GMA is impressive and in the best tradition of latter-day steam – a combination of the early simplicity of the basic firebox door, reverse and regulator, and water glasses, with refinements of the mechanical stoker and numerous additional gauges.**

304 305

306

307

307. **A steam show with few equals – the annual apple rush on the Cape Town-Elgin section over Sir Lowry's Pass. During the last full year of steam in April 1975, a 14CRB assists a GEA in the forest above Elgin.**

308. **For many years Garratts were the rule along the Langeberg, on the old New Cape Central Railway, but on every second or third Saturday the Riversdale shunting engine, to and from washout, worked the Worcester-Riversdale day passenger. Near Voorhuis, a 15BR tackles the 1-in-40 grade towards Swellendam.**

309. **For the passengers of the 'Sunset Limited' special in April 1979, a daytime run through Toorwaterpoort was a rare treat, as normal passenger trains run at night. Powering the train is a spotless 19D, one of the type which were the sole power here for more than 30 years.**

310 (Following page). **Against a backdrop of the clearly-limned Swartberg, capped with an icing of clouds, an SAR 24th Class and a Class 19D, their brightwork gleaming in the sun, haul the 'Sunset Limited' into a gentle curve on the line between Snyberg and Barandas.**

308 309

311
313

311. SAR's 'Drakensberg' no longer runs to Cape Town and on May 28, 1978, condensing 4-8-4 No. 3462 hauled the final southbound run of this, the old Blue Train. Of 90 condensing 4-8-4s only one remains – kept in service for steam specials.
312. The Karoo in summer, with an unusual covering of green after heavy rains, provided the setting for a northbound goods near Riem, with a condensing 4-8-4 in charge.
313. With smoke laid on for the photographer, double-slotted 25NCs bring the Table Bay-Witwatersrand container train up the bank from Modder River towards Kimberley. Though 25 years old these fine machines invariably run the 28 days between washouts without repairs and their overall running costs are about one third those of new diesels.
314. In the vanguard of the steam renaissance: No. 2644 was once a normal 19D and, in fact, a notoriously bad steamer. Refined by designer Wardale, she shows a decided improvement over her sister engines. Distinctive deflectors to the fore, she crosses the Modder River at Perdeberg, Orange Free State.

312
314

315. The Avontuur line's famous 'Apple-Express' runs each Saturday from July to January; here, as a short special run in conjunction with the 'Sunset Limited', it follows the scenic Gamtoos River, NG15 looking jewel-like as it hauls carriages decked in their 'Apple' colours.
316. In spite of the arid climate, snow falls regularly on the Lootsberg Pass. A light fall had not yet melted when this Class 24 worked up the northern slope of the pass in July 1975.
317. The Langkloof apple season results in intensive steam working for three months of the year, several extra trains running from Assegaaibos up the valley to collect freshly-packed fruit each day. In April 1980, an NG15 works loads up the grade from Misgund as the sun starts to set.
318. Southbound 15Fs have left Glen on their way up the last 1-in-100 before Bloemfontein, in April 1971.

315 316

317 318

319. **Natal was the home of Garratts, with Pietermaritzburg-Greytown line the particular stamping ground of doubleheaded Garratts. A lash-up of two GMAs, exerting a tractive effort of nearly 64 000 kg (140 000 lb) was needed to move 720 tonnes from Albert Falls up the 1-in-30 uncompensated grade (equivalent to 1-in-22 on the 100 m-radius curves).**

320. **Before the advent of the GMAs in the 1950s, the largest group of Garratts were the 65 GFs dating from the 1920s. One of their last SAR duties was the Donnybrook-Franklin pick-up, seen near Creighton on a frosty morning in 1974.**

321. **A line with more curves than straights was a 'natural' for the Garratt. In later years the Eshowe branch line in Zululand was the home of several GOs, but with dieselization of the North Coast in 1975, the 25 members of this class became redundant and were stored, though they still have plenty of life in them.**

319

320 321

322. South African industrial locomotives fall into two categories – those bought second-hand from SAR and its predecessors, and those built for industrial service. One of the earliest second-hand purchases was ex-NGR No. 13 bought by the Electricity Supply Commission and named 'Kitson' after the firm which built her. With more than a century under her wheels, she is still very much in service. More than 20 former SAR classes were still active on private lines in 1980.

323. Technically not industrial, since the railway it operated on was classed a 'public' railway, 'Clara' was built for the Namaqualand Copper Company's 175-km line from Port Nolloth to Okiep. It was the first 2′6″ (760 mm) gauge railway in the region, greatly influencing the choice of 3′6″ for South Africa's later railways.

324. One of two active 7th Class 4-8-0s in 1980, Witbank Consolidated Colliery No. 2 operates on this recently opened mine, which is 80 years the junior of this 1896 product of Sharp Stewart and Co.

325. Renishaw Sugar Estates operated this Bagnall Meyer, one of several examples of this unusual type which worked in South Africa.

326. Amongst the last active Dübs A 4-8-2s was this veteran at Grootvlei Proprietary Mine, shunting in the arboreal setting of the mine stores.

327. SAR designed, but built for industrial use. Eight unsuperheated equivalents to Hendrie's Class 12A 4-8-2 were built by North British Locomotive between 1948 and 1955, and Witbank Colliery has two, painted blue and spotlessly maintained.

328. Numerous engines were built to the first NGR tender design between 1904 and 1908. The 4-8-0s, which later became SAR Class 1, are good colliery workers though now nearing the end of their track. Less than five remain in service, with two operating at the Greenside colliery of Apex Mine.

329. 'King Kong', a former SAR S Class 0-8-0 heads the mine passenger trains at Grootvlei Proprietary Mine, replacing diesel powered buses in a move to beat the oil crisis.

322 324 326 328
323 325 327
329

G.V.P.M. LTD.

WITBANK COLLIERY
No4

G.V.P.M. LTD.

330

331

BIBLIOGRAPHY

Abbott, R.A.S. *The Fairlie locomotive.* David & Charles, 1970
Baker, A.C. and T.D.A. Civil. *Bagnalls of Stafford.* Oakwood Press
Brant, E.D. *Railways of North Africa.* David & Charles, 1971
Bulpin, T.V. *Islands in a Forgotten Sea.* Books of Africa, 1958
Campbell, E.D. *The Birth and Development of the Natal Railways.* Shuter & Shooter, 1951
Croxton, A.H. *Railways of Rhodesia.* David & Charles, 1973
Davies, W.J.K. *Light Railways.* Ian Allan, 1964
Day, J.R. *Railways of Southern Africa.* Arthur Barker, 1963
Day, J.R. *Railways of Northern Africa.* Arthur Barker, 1964
Durrant, A.E. *The Garratt Locomotive.* David & Charles, 1969
Durrant, A.E. *The Mallet Locomotive.* David & Charles, 1974
Durrant, A.E., A.A. Jorgensen and C.P. Lewis. *Steam on the Veld.* Ian Allan, 1972
Espitalier, T.J. and W.A.J. Day. *The Locomotive in South Africa.* In: *South African Railways and Harbours Magazine,* 1943-47
Fricke, K., R. Bude and M. Murray. *O & K Steam Locomotives.* Arley Hall Publications, 1978
Gatti, G. *Le Ferrovie Coloniali Italiane.* Edizioni GRAF, 1975
Hardy, R. *The Iron Snake.* Collins, 1965 (About Uganda Railway)
Holland, D.F. *Steam Locomotives of the South African Railways.* 2 vols. David & Charles, 1971
Kalla-Bishop, P.M. *Going Great Western to Africa and Italy.* In: *Trains and Railways.* Vol. 2 no. 9, 1974
Kalla-Bishop, P.M. *Wartime North African Railways.* In: *Trains and Railways.* Vol. 1 no. 10, 1973
Lewis, C.P. and A.A. Jorgensen. *The Great Steam Trek.* Struik, 1978
Macmillan, Allister. *Mauritius Illustrated.* W.H. & L. Collingridge, 1914
MacKay, J.C. *Light Railways.* Crosby Lockwood & Son, London, 1896
Messerschmidt, W. *Zahnradbahnen gestern-heute-in aller welt.* Franck'sche Verlagshandlung, Stuttgart, 1972
Miller, C. *The Lunatic Express.* MacDonald, 1972. (About Uganda Railway)
Moir, S.M. *Twenty-four Inches Apart.* Oakwood Press, 1963. (About South African narrow gauge)
Moir, S.M. and H.T. Crittenden. *Namib Narrow Gauge.* Oakwood Press, c.1957
Patience, K. *Steam in East Africa.* Heinemann, 1976
Pauling G. *The Chronicles of a Contractor.* Books of Rhodesia, 1969
Ransome-Wallis, P. *On Railways, At Home and Abroad.* Batchworth Press, 1951
Patterson, J.H. *The Man-Eaters of Tsavo.* Fontana, 1973
Pereira de Lima, A. *História dos Caminhos de Ferro de Moçambique.* Ediçéo Ceni, 1971
R.C.T.S. *The Standard-gauge Locomotives of the Egyptian State Railways and the Palestine Railways, 1942-45.*
Raemar, R. *Steam Locomotives of the East African Railways.* David & Charles, 1974
Reed, B., editor. *Locomotives in Profile.* Doubleday, 1971
Rowledge, J.W.P. *Heavy Goods Engines of the War Department.* Springmead Railway Books, 1977
Schroeter, H. *Die Eisenbahnen der Ehemaligen Deutschen Schutzebegiete Afrika und ihre Fahrzeuge.* Wirkerswissenschaft Lehrmittel Gesellschaft Mbh., 1961
Small, C.S. *Far Wheels.* Cleaver Hulme; Simmons Boardman, 1959
Talbot, E. *Steam from Kenya to the Cape.* Continental Railway Circle, 1975
Varian, H.F. *Some African Milestones.* Books of Rhodesia, 1973. (About building of lines in Rhodesia, Moçambique, Angola and Congo)
Vuillet, G. *Railway Reminiscences of Three Continents.* Nelson, 1968
Weinthal, L., editor. *The Story of the Cape to Cairo Railway and River Route from 1887 to 1922.* Pioneer, c.1922
Whitehouse, P.B. and Peter Allen, *Narrow Gauge the World Over.* Ian Allen, 1976
Wiener, L. *Articulated Locomotives.* Constable & Co., 1930
Wiener, L. *Les chemins de fer coloniaux de l'Afrique.* Goemare; Dunod, 1930
Wiener, L. *L'Egypte et ses chemins de fer.* Weissenbruch, Brussels, 1932.

Official diagram books: EAR&H, SAR, RR, CFM, CFB, BCK, ESR.

Locomotive builders' catalogues:
Alco
Ansaldo
Armstrong Whitworth
Baldwin
Beyer Peacock
Borsig
Breda
Brush
Haine St. Pierre
Henschel
Krauss-Maffei
La Meuse
Robert Stephenson
St. Léonard
Tubize
Vulcan Foundry

Periodicals:
Beyer Peacock Quarterly Review
Continental Railway Journal
Industrial Railway Record
The Locomotive Magazine
The Narrow Gauge
Railway Directory and Yearbook
The Railway Gazette International
Rhodesia Railway Record
S A Rail
Train
Trains & Railways

330. Sunrise at Bronkhorstspruit. A pair of 15CA 4-8-2s storm up the grade with the daily international train from Lourenço Marques (now Maputo) during 1975. Electric locomotives now take this train from the border at Komatipoort to Pretoria.

331. October and November each year reveals the lineside jacaranda trees in full bloom at Cullinan. Trains on this branch normally have 15CAs – their last regular line working – but during 1976 No. 1791, an un-reboilered 15A was a special attraction. Even in this late day of SAR steam, special engine workings keep local enthusiasts busy.

PHOTOGRAPHIC CREDITS

(Numbers refer to pictures)

J. Adams collection 130, 131
Sir Peter Allen 59
J. Allerton 105, 107, 114, 231, 239, 262
P.F. Bagshawe, 129, 133, 134, 141, 204-207, 256, 269-275
Hugh Ballantyne 3, 94, 98, 104, 111
P.J. Bawcutt 41-44, 48, 51, 53, 62, 72-74, 86
A.E. Durrant 24, 75, 99, 106, 116, 160, 198-199, 201, 222-223, 258, 261-263, 268, 292, 304, 307, 311, 315, 321, 324, 326-329, 331
A.E. Durrant collection 11-12, 15, 25-29, 31, 46-47, 58, 60, 66-67, 71, 77, 92-93, 155-156, 166-172, 192-193, 245, 247-253
Christine Durrant 267
K.W. Clingan collection 34
F. Fenino 13 (collection *La Vie du Rail*)
N. Fields 301
F.G. Garrison 299
R. Guenin 179 (collection *Railway Gazette*)
J. Hall 113, 218, 225, 228-229, 236-238
N. Huxtable 4, 64-65, 68-70, 102, 109-110
A.A. Jorgensen 2, 5, 159, 196, 209, 211, 213-214, 219-221, 226-227, 230, 233, 235, 241-242, 260, 265-266, 291, 293, 295, 298, 302-303, 306, 308, 317, 320, 322-323, 325
A.A. Jorgensen collection 91, 121, 128, 135, 137, 142, 165, 174
W.H.C. Kelland 50, 57, 61 (collection Bournemouth Railway Club)
G. Kempis 278, 281, 282, 284
Kimberley Public Library collection 259
R.A. Kingsford-Smith 69, 90, 108, 115
La Vie du Rail collection 6-10, 16, 33, 143-148, 150-152, 154, 157-158
C.P. Lewis 1, 195, 197, 200, 202-203, 208, 210, 212, 215, 217, 224, 232, 234, 240, 264, 276, 294, 296-297, 305, 309, 310, 312-314, 316, 318-319, 330
L.G. Marshall 23
S.M. Moir collection 280
North Western Museum (England) 182
K. Patience collection 136
H. Pearce 17, 19, 22
J.H. Price collection 14, 153
Railway Gazette collection 52, 175, 177, 180
Dr. P. Ransome-Wallis 56, 138, 163, 300
R F F S A (Brazilian Railways) 49
C.E. Rickwood 244, 246
B. Roberts 95-97, 100, 101, 112, 117, 162, 216
S. Robertson 285
A.M.S. Russell 30, 32
C. Salermon-Bosch collection 140
C.S. Small 76, 78-80, 87, 89, 119-120, 122-127, 254-255, 257
Dr. G. Smith 118
S. Smyser 139
South African Railways 277, 279, 286-290
The late Tony Spit collection 283
R.E. Tustin 63 (P.B. Whitehouse collection)
P.B. Whitehouse 54, 55
P.B. Whitehouse collection 243
W.C. Williams (Beyer Peacock) 18
J. Wilson 45
Jeremy Wiseman 20-21, 35-40, 81-85, 88, 103, 132, 149, 161, 164, 173, 176, 178, 181, 183-191, 194

INDEX